Fire on the Mountain

Also by Theo Maehr:

Wild Whispers
When Nature Speaks, Listen
Illustrated by Sonja Lokensgard

Fire on the Mountain

Living with Wildfire in the Santa Lucia Mountains of Big Sur

Theo Maehr

Keep the Wild
Big Sur, California

Copyright © 2024 Theo Maehr

All Rights Reserved

Published by Keep the Wild
Big Sur, California

www.theearthsteward.com

ISBN: 979-8-218-40776-6

This book and its contents are protected under International and Federal Copyright Laws and Treaties. Except for the quotation of short passages for the purpose of criticism or review, no part of this book may be reproduced or transmitted in any form or by any means, electronic or mechanical, including photocopying, recording, or by any information storage and retrieval system, without express written permission from the author.

Book design and production
by Lucky Valley Press
www.luckyvalleypress.com

Printed in the USA on acid-free paper

Dedication

I would like to dedicate this book to the wildlife we share this land and this earth with – all those living beings whose silent voices need to be heard and whose presence makes living on this planet beautiful, rich, and diverse.

Fire on the Mountain is also dedicated to my mother, without whose behind-the-scenes guidance and support I never could have created my home on the mountain. She it was who set me and my brothers free to explore wild places, and supported us in our dreams and aspirations. She knew me well enough to create opportunities fostering my love of nature and my quest for living within it. My early life was comprised of seemingly normal experiences, but compounded sequentially they set me on a path of oneness with the earth. From being able to raise and nurture wild animals in my room at home, to living as a ranch hand in the wilds of Montana, my mom opened doors for me to pass through—inspiring dreams held deeply within—to understand fully the world and the life around me and to realize my connection to it all.

Contents

Foreword by Lee Klinger v

1. Calamity Knocks at My Door 1
2. The Mountain Property 9
3. The Soberanes Fire Arrives 17
4. The Fire Saga Continues 33
5. A Death and Destruction 41
6. Another Vivid Dream 51
7. Mid-Coast Fire Brigade 53
8. I Leave the Mountain 59
9. Returning Home 69
10. Surrender 77
11. Living with the Aftermath 81
12. Devastating Winter Storms 85
13. Nature Delivers a Special Treat 89
14. Culvert Construction 95
15. NRCS Creates a Forest Management Grant 101
16. Living into the Present 111
17. House and Land Performance 117
18. Making Your Land Fire-Safe 123

Acknowledgements 133

About the Author 135

Photo taken on July 22, 2016 in the early evening, looking south from the parking lot of the Sunset Auditorium in Carmel toward Soberanes Canyon and Big Sur. The Soberanes Fire had begun that morning.

Foreword

by Lee Klinger

Soon after arriving in Big Sur in 2005, I had the fortune to meet Theo Maehr, a bright, strong, and caring man with whom I immediately felt a kinship. His youthful appearance seemed at odds with the many years of life experiences he shared with me—farming pursuits in Ohio, studies at Stanford, teaching at a Waldorf school, and homesteading in the Big Sur highlands.

While many skillful locals have succeeded in finding ways to survive and flourish in this challenging mountain landscape, Theo has done it by being a true steward of the land. I am reminded of this every time I receive jars of his honey, olives, and olive oil, or fresh fruit from his orchard, or a delivery of madrone firewood for heating my home.

When I first heard, in real time, about Theo's heroic efforts during the Soberanes fire of 2016, I felt simultaneously in awe at the struggles he had to endure, yet unsurprised to find out he was well-prepared. His fire-resistant infrastructure, massive rainwater storage system, years of firefighting experience, and nurturing connections to the land and his neighbors were all in anticipation of such an event.

Now, after reading here the full account of Theo's real-life experiences during the Soberanes fire, I find myself deeply appreciative of how his focused efforts were supported by his community of family, neighbors, and local fire fighters. What I find particularly fascinating in this work is the careful attention to details of the events and people involved, told chronologically in a way that made me feel the passage of time on the mountain, often weaving together accounts involving simultaneous nearby fire fronts in the Long Ridge neighborhood.

The Soberanes fire was just one of more than a dozen major wildfires that have burned in this region over the past few decades. I remember the rare summer storm event in June of 2008 when I witnessed a dry lightning strike on a nearby mountain that ignited a small forest fire, which eventually merged with several other lightning-started fires to form the huge Basin Complex fire. I spent more than two weeks preparing the land around my home in case the fires arrived, sometimes having to extinguish glowing embers that fell nearby.

My home and surrounding forests escaped the Basin Complex fire, but others were not so fortunate. Then came the Pfeiffer fire of 2013, and it was my turn to feel the heat.

Anyone who has lived in Big Sur for the past few decades has a story of how a nearby wildfire has affected them either directly, as in my case by losing their home, or indirectly by damage to roads and other important infrastructure.

Still, every fire is a lesson from nature and those who can learn and adapt will recover for the better. For my part, besides having better fire clearance around my new home, I have found a livelihood in helping others create fire safe forests on their land.

This is all to emphasize the importance of wildfire preparedness in Big Sur. It's not a question of "if" the wildfire arrives, it's a question of "when." This book demonstrates that the foresight of knowing the inevitability of wildfire is essential for one's survival in the forests of Big Sur and beyond.

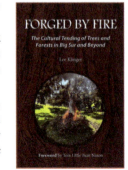

Lee Klinger, Ph.D., is an Independent Scientist and Consultant in Big Sur, CA currently working with the Department of Natural Resources of the Esselen Tribe of Monterey County, and with the Mutsun Costanoan leaders at Indian Canyon Nation.

Since 2005 he has served as the director of Sudden Oak Life, a movement aimed at applying fire mimicry practices to address the problems of forest decline and severe wildfires in California.

He has more than forty years of experience in forestry, plant and soil ecology, atmospheric chemistry, earth system science, and nature photography. Lee is the author of *Forged by Fire: The Cultural Tending of Trees and Forests in Big Sur and Beyond*.

1

Calamity Knocks at My Door

Times of great calamity and confusion have been productive for the greatest minds. The purest ore is produced from the hottest furnace. The brightest thunderbolt is elicited from the darkest storm.
– Charles Caleb Colton

Necessity may well be called the mother of invention but calamity is the test of integrity. When any calamity is suffered, the first thing to be remembered is, how much has been escaped.
– Samuel Johnson

I was not fully prepared when wildfire arrived. How could I possibly know the intense feelings I would experience when all I had worked so hard to create would suddenly become threatened? When fire, that powerful force of nature, met me, my only hope was to remain present and ready to respond in whatever way necessary.

I had spent years preparing for the wildfire I knew was inevitable in our locality. Along the Pacific Coast in Big Sur, fire is a part of the landscape. On the morning of July 22, 2016, when this fire began, I had been a firefighter for Mid-Coast Fire since 2010. A few hours after the fire started Mid-Coast Fire firefighters were called to respond to a wildland fire in Soberanes canyon. Unfortunately, I had awakened with a 102-degree fever.

Despite feeling horrible at 8am, I rallied to meet with Pete, the owner of Peninsula Septic Tank Service, to finalize a plan to have my septic system pumped out the next Wednesday. We were just saying goodbye when my pager went off calling me to respond to a wildfire in Soberanes Canyon. Amazingly, I was parked just below the fire station. I quickly said goodbye to Pete, left my car parked where it was, and headed up to the station. Little did Pete or I know that it would be almost a year before PSTS would be able to make it up to my property to pump out the system.

At MidCoast Fire Station, Norm Cotton had also shown up for the call, and Nathan Estrada was already there on duty at the station that day. Within minutes we all had our wildland fire gear donned and were in one of the fire engines on our way to the fire.

When we arrived, there were already fire crews on scene from CalFire and Carmel Highlands. Thick black smoke was rising from a spot a couple miles up Soberanes canyon. It was coming from a place I knew, near a stand of redwoods along the creek. I had always thought that place would be ideal for sleeping out, but camping was not permitted there. I imagined illegal campers negligently starting the fire. Months later, the results of investigations confirmed my suspicion.

Though Pete did not arrive in this truck to determine how to assess my septic system, his sense of humor was present.

When we radioed in from our position on Highway 1, the Incident Commander (IC) directed us to stay put and wait for further instructions.

As we sat there in the engine awaiting those instructions, we watched as three men with full backpacking gear walked out of the canyon. Though State Park personnel, sheriffs, and CalFire were standing in a circle talking no more than 30 yards away, no one did anything to question or detain the backpackers. No one seemed to notice them at all, and though I was ready to get out and do the confronting, as a firefighter who had arrived on scene, I was under the supervision of the IC. In that position I could do nothing without bringing negative attention to myself and MidCoast Fire Brigade.

Eventually the IC got back to us. His words came crisply through our radio's speaker, "We've got it handled. You are not needed. You can head back to your station."

We all shared the opinion that they actually did not have it handled. Norm responded to the IC and offered to lead firefighters up a private road that would position them above the fire, allowing for a good view of it, and providing the opportunity to create a line of defense should the fire continue to advance.

The IC declined the offer and sent the local boys home. So, in the top-down style that fire fighters learn to live by, we turned the engine around and headed south on Highway 1 toward Palo Colorado Canyon.

When we arrived at the station, I got out of my gear and lay down on the cool concrete driveway outside the engine bays. I was feeling feverishly hot on the inside, and it was already a warm July day. The concrete drew some of the heat out of my body, but I felt horrible. Eventually, realizing concrete was not the most comfortable resting surface, I made my way to my truck and drove the four miles home. As I lay down on my couch, I could hear the drone of a helicopter and a spotter plane. It sounded like the, "We've got it handled," unfortunately might not be true.

When the fire began it was about seven miles from my property. By nightfall the fire had progressed over several ridges and become unmanageable and out of control. It was heading directly toward the Palo Colorado community where I lived and had covered half the distance to it. I could see the glow of the fire from my home and felt certain it was only a matter of time before the fire would envelop my property. There were a lot of things I could do to prepare for the inevitable, but I was sick and wanted to do nothing but rest and sleep. I tried to get to work, but my body refused. I considered all the things I should be doing as I fell in and out of a restless sleep.

By Saturday morning the fire had advanced a couple more miles in my direction. My friends Mark and Suzanne had a place on Garrapata Ridge, a few ridges closer to the fire, and called to check in with me. They and a friend had been up since 6am, packing furniture and belongings into a fireproof basement. The fire was across the canyon from their place, slowly burning down the hill and making its way toward their house and property. They were worried they might lose everything on their property, and they had a lot to lose.

They purchased the property in 2003, a year before I purchased mine. On it were gardens and orchards, and a beautiful hand-built home. Since buying it, they had refurbished the house, built decks, stone walls, and stone patios. They had also planted hundreds of fruit-bearing trees—90 olive, 90 avocado and a variety of other food-bearing trees. I had helped to install garden beds, worked on the irrigation system, and done a variety of small building projects. To hear that the fire was close enough to be threatening everything was alarming.

Though I wanted to get out and start preparing, every little thing I did took its toll. I still had a fever and became exhausted easily, could not catch my breath, and just wanted to rest. By early afternoon, the entire canyon was under mandatory evacuation. Mark called to let me know they had done all they could and were leaving. He wished me luck and said he would telephone soon.

I lay down again, wishing I could rally and prepare for the fire, but I still had no energy. I felt like I was in some weird nightmare that was not going away. When I awoke again it was 4pm and there were visible flames on the ridge to the north—1.5 miles away.

There were things I needed to do. I had experienced enough fires as a firefighter for Mid-Coast Fire Brigade to know that in a matter of hours it would be on my property.

I had not eaten anything all day and timidly sipped warm water. My mouth had that nasty taste that comes with a fever. The water helped a little. I reached for a container of chocolate covered espresso beans, and downed a handful hoping they would give me a little boost.

I walked outside. The air was still clear around the ridgetop. The mild breeze was blowing the billowing black smoke northward, but not deterring the fire's advance.

I got started. I moved flammable materials away from buildings. I hitched a truck to an already filled 500-gallon water trailer equipped with a water pump and fire hose. I hitched my horse trailer to another truck in case I needed to evacuate the horses. I rolled out two fire hoses with attached nozzles and connected them to fire hydrants from my water system. I sealed off open areas beneath a couple buildings. I moved clothing I wanted to save into my concrete kitchen, and valuables into the cold room/fire shelter (constructed next to and adjoining the kitchen).

Pat Pallastrini, my closest neighbor, came over trying to persuade me to leave. I told him I was staying, assuring him I would be safe given the cold room I had built into the earth, and the 500-gallon water trailer with pump and fire hose. We looked north across the canyon to the fire on the ridge. It did not look good. He again expressed his concern, but

I had made up my mind to stay. Pat offered me anything I needed from his place, water from his tanks or food from his house—a generous and thoughtful offer.

After a hug and well wishes, Pat left and I returned to preparing for the fire.

As I worked, I watched as vehicles rolled down off our ridge road and out of sight. I was tempted to get in a vehicle and drive away from this far too real catastrophe. I saw my friend and neighbor Zach driving down the hill. Zach was a rugged mountain man, and not to be frightened by a fire. Earlier, I had thought that of all people at least Zach will stay on the ridge. I again thought about leaving, but it did not seem like an option. I reasoned that the only way to save my home was to stay, and if need be, duck into the cold room that doubled as a fire shelter.

The fire was raging below me in Rocky Creek. From my vantage point, I could see that it was moving in many directions at once.

I felt bewildered. What was I doing? Why had I stayed? How did I get this crazy idea that I could battle this fire to save my home? The fire was intense, and half a mile below me had enveloped both sides of Long Ridge Road—the only escape route.

I thought again about the cold room/fire shelter, which I built in 2011. When I first moved onto the property, I realized it would be essential to have a room that could function as a fire shelter. All sides were buried except the one with the entrance. The entire room was built of concrete. The side with the entrance was concrete as well with a steel door sealed with rubber gaskets. Within it there was water, bulk foods and two filled SCUBA tanks. I felt confident that even if I was unable to save my structures, I could ride out the fire in that room. The SCUBA tanks would provide hours of clean air and create a positive pressure within the room to keep the smoke out.

I fondly surveyed my property as I prepared for the fire. It had been a little over ten years since I first began construction of my home. I had started with a raw piece of land and had put all my resources into creating my vision of home. I had imagined, designed, collected all sorts of things, planted, and built. The result was an off the grid living situation with multiple buildings, an edible landscape providing an abundance of food, and a lush environment nourishing and supporting a diverse and abundant wildlife population. It was a beautiful place.

Thinking about all that, I felt renewed determination to do everything possible to save it.

The sun set. I headed inside briefly and ate a little. Working had given me an appetite. After a bit of nourishment, I headed back out to clear leaves and debris off roofs and out of gutters.

I climbed onto my shop roof. I thought how strange it was to be alone on the ridge. From my vantage point on the roof, I could see the fire roaring through Rocky Creek. It was a fast-moving fire and its intensity frightening. I wondered out loud to myself, "What am I thinking? I'm crazy thinking I can survive through this." My cold room was some comfort, but it had never been put to the test and I did not know whether it would really work. If the fire continued to behave chaotically, I might have a situation I could not control when it got to my place. I knew I might lose everything and even be struggling to save my life.

In the last of the evening light, Christian, another close neighbor, drove up my driveway to check in with me. I wasn't alone on the ridge! What a relief to realize a friend had stayed. I continued sweeping leaves off as we talked. He climbed up and began to help me. July in a madrone forest is like Fall in most forests. The trees shed old leaves during this month and create a huge litter of leaves over the ground, onto roofs, and into gutters. It was essential to get these leaves off the roof of any building I wanted to save. A single ember could ignite a pile of leaves and once started, a fire could easily go from burning organic matter to burning a building.

We were both anxious and keenly aware of the danger the fire posed for our properties and our lives. Christian decided to drive down the road to see what the fire was doing. He returned with the news that the fire was moving quickly up the northwest side of the ridge, about ½ mile away. He joined me to sweep off the house roof. His help and company were comforting. By the time we were done cleaning the house roof and gutters, it was getting too dark to see without a light source. Christian again left to go check on the fire. He said if he drove by and honked, I better be ready. The honk would mean the fire was coming fast. A little while later I watched him drive past. He did not honk.

I climbed off the roof and again went inside. I ate a little more food, drank some tea, and decided to take a bath and refresh myself before the inevitable. As I dried off, I watched from my house as huge flames shot high into the sky indicating a house was probably on fire. The fire slackened briefly before becoming huge again as the place across from mine burned. The fire was on its way, and I would soon need to try to stop it from coming further up the ridge.

I wondered about the untimely fever I was experiencing. Why was I having a fever at such an inopportune time? I would have liked working more actively on our ridge top and to have shown up to at least try to save the two houses that burned, but with the condition I was in, I simply did not have the energy to do so.

I did not have time to tend to my barn, which was closest to the fire. As I pulled on my fire boots, I thought the chances of the barn making it through this fire were slim.

I looked out the windows and could see the fire coming down the hillside onto my property to the west. It was time to go meet the fire.

I went outside to an old 4Runner and loaded into it a couple full water bottles for drinking, and a shovel. I drove down my driveway toward the fire and parked about 150 feet away. Grabbing the shovel, I got out and walked toward the fire. As I did, a state of calm came over me. I found myself fully in the present moment with senses full of input from the dynamically burning landscape.

The outcome over the next hours was completely uncertain. Would I lose everything I had worked so hard to create, including perhaps my life, or would I survive with my place intact?

Aerial view of the Soberanes Fire. Theo's property is engulfed in smoke.

Above: View from property toward the Pacific
Below: View from property toward Pico Blanco

2

The Mountain Property

I bought a 44-acre property in March 2005, and began building a house on it in August of that year. By then I had finished the interior of a cabin built by the previous owner, and a small shop in its entirety, out of recycled building materials. I built the shop to accommodate larger woodworking machines—table saw, planer, jointer, drill press.

It took eight months of focused work to bring the house to a habitable state. When I moved in, there were still many things to finish and even at the time of this writing some interior trim work remains to be done.

In 2009 I started building an 18'x30' two-story barn down the hill from the house, and at the time of the Soberanes Fire, I was still working to finish the interior. Originally, I planned to make the second floor a hay loft, but once in the flow of building realized that if I raised the roof 5 more feet, I could make the second floor into a meeting/art and yoga space. I raised the roof. The views out the windows toward the southwest were stunning, looking down toward Bixby Canyon and the Pacific.

In 2011 I built a 9'x10' cold room/fire shelter and an adjoining 18'x10' kitchen. The cold room/fire shelter was buried on all sides except the side with the entryway, and the kitchen was built into and partially covered with earth. I had envisioned building this structure when I first bought the property. I knew having a fire shelter was essential to living safely in the higher altitude, Big Sur landscape. At 2700', the land was hot and dry in the summer, and the moisture content in the vegetation became very low, making it extremely prone to wildfires. The best chance to save structures from burning during a fire event was to be present on the land. The fire shelter would provide the safe refuge to be able to do that.

Rain collection was the water source. I continued to add tanks until I had 60,000 gallons of water storage—enough to make it through the dry summer months and to the time in the fall when the rains almost always came. In 2014, the rains had still not begun by late December and only 8000 gallons remained. The day before Christmas I picked up a 500-gallon water trailer to transport water from a neighbor's system into mine. Though I hoped the rains would come, this was preparation for the worst—that the drought conditions would continue. A few days after Christmas the rains for that season began, but not before I had transported a few thousand gallons of water into the storage tanks.

The good news was that the water trailer increased the chances of being able to defend the property during a fire event. Equipped with a 500-gallon water tank, a water pump, and a 50-foot length of 1.5 inch fire hose, it was like having a mini fire engine.

Close to the house, I had planted hundreds of food bearing trees (apricots, peaches, plums, pluots, figs, almonds, apples, persimmons, pears, oranges, lemons, limes, grapefruit, cara-cara, pomegranates and olives), watering them weekly during the dry months for the first 5 years and until they became established. There were also a number of raised garden beds.

I had put a great amount into creating home on the mountain, and the thought of losing it all did not sit well with me. Knowing it was a precarious existence to live in the back country of Big Sur, it made sense to make the area around my buildings, orchards and gardens more fire safe and less likely to burn intensely if there was a fire. Every year countless hours were spent clearing and burning brush during the winter months when brush burning was permitted.

Though in the past there had been fires that threatened to burn up to and through Long Ridge, the Soberanes Fire was the first to actually do so.

With the certainty of the Soberanes fire coming onto my property, it would soon be clear whether all I had done to prepare for a fire would work out to keep my buildings and landscaping from burning.

Above: View through the madrones to the Pacific

Right: Location of property with Bixby Bridge in the foreground

Above: The enchanting forest   *Below*: Horses

Above: Raven   *Below*: View from the Treehouse

The Main House

East Side

North side

West side

Outside Kitchen with
attached cold room

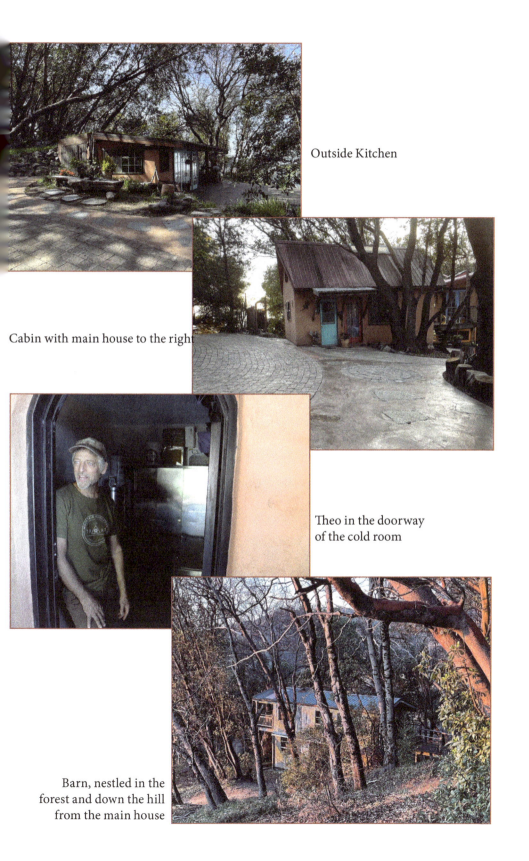

Outside Kitchen

Cabin with main house to the right

Theo in the doorway of the cold room

Barn, nestled in the forest and down the hill from the main house

The Soberanes Fire making its way across the Big Sur landscape

The Soberanes Fire on its way

3

The Soberanes Fire Arrives

Shovel in hand and faced with the possibility of losing everything I had worked to create, I discovered within me a state beyond any fear of loss. I found myself in a state of surrender and acceptance—present and in the moment.

There was no room for thinking about anything other than what was directly in front of me—the landscape, the fire, and my relationship to them. The situation called for my undivided attention.

The fire had made its way down to a saddle like area—an area I had already extensively cleared. As I walked over to the fire, I felt as if I was meeting someone, and my introduction was not an unpleasant one. I was aware of the bright, energetic brilliance of the fire and felt there was an underlying intelligence within it. It did not seem destructive or angry.

The area where I first encountered the fire was at the top of a slight rise. Amazingly, the fire was moving slowly through leaves and inconsequential brush and was having difficulty progressing in the open areas free of vegetation, between the remaining manzanita and chamise. Further to the southwest and down a steep hill the fire was burning in an area of heavy brush, where I had not cleared. Flames were 10-15 feet high and the fire was burning intensely. It was backing down toward a fire road below it.

The active edge of the fire was bound on either side by roads, leaving me about 150 yards I needed to work. I knew I could not stop the fire burning through the heavy brush. The only thing that made sense was to begin with the smallest fire, on the saddle, closest to Long Ridge Road. If I could extinguish the fire there, then I could work my way to the south and the area burning more intensely.

I thought to myself, *Just focus on what is before you, you'll know what to do.*

As I began to work, I did not feel at odds with the fire. It did not seem like fighting as much as it felt like collaborating. I worked to put the fire out in the places where it was smallest. There were times when I laid down on the cool earth next to the fire to get away from the smoke and to rest. Being feverish made physical exertion exhausting. I felt like I was climbing a mountain in high elevation. I couldn't keep my breath. I could work, but I tired quickly and had to will myself to keep going.

Slowly, I stopped the fire's progression on the saddle. When I was confident it was extinguished there, I dropped down to the fire that had been the most active toward the south, and arrived during a time when there were few flames actively burning. The flames had been high and the fire intense as it came up the hill, but once it reached a fire road, it had no fuel to continue to burn. Along the edge of that burned area I now descended, extinguishing the small flames as I proceeded. Some I simply stomped out, while others took a small amount of shovel work—throwing dirt onto the flames to smother them. The amazing thing was that this area had been burning with 10-15 foot flames just a short time before. Now, I was walking along its burning edge and extinguishing it easily by knocking the flaming tips of branches to the ground and shoveling earth on the burning duff—the decaying matter covering the ground under shrubs and trees.

I had seen this behavior on other fires. The fire will make a rush up a hill burning a 10-20-foot-wide swath of vegetation. Once at the top, it will slowly back down the hillside, before making another uphill rush.

Over the course of seven hours, using only a shovel, I stopped the fire's forward progress for that 150-yard stretch. As the hours passed, I made my way back and forth along the fire line, continuing to put out little fires and increasing the width of a line of bare mineral soil which was keeping the fire from progressing. Eventually, when the active fires were out, I went up to the house and returned with the water trailer, using the 500 gallons to extinguish the fire where it was still burning in manzanita root balls. Surprisingly, no matter how much water I drenched over some of the root balls, they would not go out completely.

After having used the 500 gallons, I walked over the entire line again making sure any active fire was contained in an already burned area. There was one small plume of smoke coming from the edge of a brushy wash. To get to it would mean pushing my way through almost impenetrable brush. I left it, feeling too exhausted to make the effort, while knowing I might regret that decision later.

Darkness was giving way to light. Morning was coming. By now eight hours had passed. The fire was at a standstill. There were still some smoldering root balls, but nothing was threatening to burn further up the ridge toward my buildings, gardens, and orchards, or the rest of the neighborhood.

By the time dawn arrived, I thought, *At least I've made it through the night*. During those dark hours, I extinguished the fire that had come onto the western edge of my property using nothing except a shovel. Exhausted and elated, I lay down on the cool earth. I was feverish. Working for hours near the flaming brush had not diminished my elevated temperature. I did however feel jubilant. I had made it through the night and my buildings were still intact. It felt like a miracle.

Knowing I might need to use the water trailer again, I filled it via a hydrant I had installed along the edge of the main road up Long Ridge. Afterwards, I went back to my house, drank some tea, and sat for a few minutes. I checked my phones and saw that I had received over 20 texts and numerous calls from concerned friends. My body was so awake, even though I had not slept the entire night. I still felt on high alert.

Prepping my land for fire

Ever since the Basin Fire in 2008, when I watched the fire stop on the ridge a mile to the south of my property, I became more focused on fire preparedness. After the 2013 Pfeiffer fire I knew it was only a matter of time before fire would one day threaten my property.

Over the ten-year period living on my property, I had diligently done many things with the knowledge that wildfire was the most serious threat I faced living on the ridge. Most of the California landscape is predisposed to fire events. The flora and fauna have adapted to periodic fires and is healthier when there are regular burns. The indigenous people living in Big Sur, California had reportedly used fire to manage the landscape, keeping it vibrant, and sparing it from major fire events. Our current building practices, in which we attempt to build 'permanent' structures, does not easily allow for periodic healthy burns. Because we now suppress fires, flammable fuel loads build up. When fire inevitably does make an appearance, the fires tend to be uncontrollable and catastrophic. These unnaturally large, hot fires destroy the ecosystems that depend on regular burns, making it difficult for the life there to make a reasonable comeback.

Why do we look for such unnatural permanence in the ever-changing phenomenal world, especially in ecosystems that change so dramatically? The nomadic cultures, once spread across the planet, lived in harmony with the rhythms of the natural world and the continual process of creation and destruction. But our current world order has us attempting to create structures that will endure through the centuries. The industries built up to support such an attempt consume our earth's resources at an unconsciously exhausting rate, and with that orientation are obliterating life as we have known it on the planet.

To keep structures and landscaping installations from burning on any piece of land, it is critical to mimic the effects of fire over the landscape. In Big Sur, every winter, brush clearing, and chipping or burning, seems the best course of action to take. This not only creates a defensible zone around structures, orchards, and gardens, but brings health and vitality to the living landscape, supporting rare native plants and wildlife that cannot exist without periodic fires and cleared land.

By the time of the Soberanes Fire, I had done my best to manage the landscape closest to my buildings in such a way.

In addition to maintaining the landscape by clearing and burning brush, when I installed my water system, I plumbed in fire hydrants, and collected fire hoses and nozzles. Having the 500-gallon water trailer added another layer of security.

After refreshing myself at the house, I hopped back into the 4Runner to drive the ridge to see what everything looked like.

As I drove further up Long Ridge it was clear the fire had not made its way past the line I had created. Everything was still and quiet in the morning. I drove all the way to the end of Long Ridge and down to the Jensen's place. I drove back the other direction and stopped to clear fallen branches out of the road. A voice beckoned, "Who is that?"

I responded with the same, "Who is that?"

Jake Goetz, Mid-Coast Fire Brigade Assistant to the chief, appeared around the corner, walking up to me and pulling me into a hug. "We have been so worried about you!" he exclaimed. Cheryl, the chief, and

Jake's wife came up and hugged me as well. They knew I had stayed on the ridge with a fever. Two fire brigade members, Andreas Bear and Brian Gorrel had called and left messages for me to come down off the mountain, but I had not answered their calls. I had not wanted to talk to anyone and be persuaded to leave the mountain. Having anyone reinforce the doubts I already had would not have been good.

"I'm fine," I said and then recounted to Jake and Cheryl the story of the night. I credited them for the work I was able to accomplish, thanking them for all the training and experience I received during my six years on the Brigade. The training and experience had given me the confidence and knowledge to deal with the fire.

My time on the Mid-Coast Fire Brigade

When I moved into a rental house in Palo Colorado Canyon in 2004, there was an active Fire Brigade in the canyon, with a fire station in the final stages of construction. During the next few years, I was too busy figuring out my own property and situation to join the brigade, but I watched with interest as the fire station was built, and slowly became acquainted with members of the Mid-Coast Fire Brigade.

In October 2007, the year after I moved into the house I had built on Long Ridge, a fire started very near the fire station. It was called the Colorado Fire. I was driving home from a day of teaching and had stopped at the mailboxes just below the fire station to retrieve my mail. The plume of smoke coming from the canyon below me was frighteningly noticeable. As I gazed at it in relative disbelief, a car with two frantic women in it came zooming up to the fire station. At the same time, Norm Cotton, a longtime fire brigade member, drove to the fire station. He parked and opened one of the bays. As he donned his fire gear, I ran up to the fire station and asked if I could come with him to help.

He replied, "I can't let you ride on the engine, but you can follow along behind me in your car and help me when we get to the fire."

So, as Norm pulled away from the station, I fell in behind him, keeping back a safe distance. We drove down the hill away from the fire station and toward the coast. Norm turned onto a narrow one-way road that accessed a number of houses. A car was coming out and let Norm go by but detained me. Once it had passed, I continued up the road. Norm was now out of sight. I rounded a bend and ahead there was active fire on both sides of the road. The flames were 3-5 feet high. I drove to within 15 feet of the flames and thought to myself that maybe it was not the best idea to drive through those flames. If I knew what I know now, I would have kept driving past them, but at that point it just seemed like too much.

I backed my car down the hill a short distance and into a turn out where it would be out of the way. I closed the windows of my car, got out, and began to run up the road thinking I could get to where Norm was more safely on foot, and not endanger my car in the process.

I had run about 50 yards up the road and was just getting to the part of the road where there were flames on either side, when the sound of an airplane quickly approaching filled my awareness. I looked up in the direction of the fast-approaching plane. It was not visible through the smoke until the last seconds, when it became clear it was a CalFire plane dropping fire retardant. The plane opened its bay and the orange retardant began raining down toward the earth directly over the place where I stood. I turned and ran a few steps to take shelter beneath a live oak next to the road. The leaves of the tree absorbed most of the retardant, though I was still hit by some of it.

I decided at that point I was not dressed properly for firefighting and did not know enough about what I was getting into, to continue up the road to where I knew Norm was fighting the fire. I hoped I might be better able to help elsewhere. In that moment I decided I did not want to be in a similar situation again. I would do what it took to join Mid-Coast Fire Brigade and be able to respond to emergency situations in my community.

The rest of that day I spent on the property of my friends, Mark and Suzanne, west of the fire. From a place on their property, I could watch the fire's activity and if it came in that direction, I could help keep it from burning their place. The problem was that I really did not know what I would do. I had no training.

The day after the Colorado fire (it had been extinguished in hours) I talked with Cheryl Goetz, Mid-Coast Fire Chief. She let me know that to become a member of the brigade I would need to do extensive training and there were no trainings scheduled yet, but that she would love to have me join.

Mid Coast Fire Brigade – jaws of life training day (Theo far left)

Theo as a firefighter for Mid Coast Fire Brigade

Water Trailer

Mid Coast Fire Brigade – *Top Row*: Leonard Lazuras, Brent Bispo, Scott Bogen, Travis Trapkus, Norm Cotton, Andreas Baer, Kerri Frangioso. *Bottom Row*: Giau Nguyen, Cheryl Goetz, Theo Maehr, Jake Goetz

Time passed. I was busy with my life and my schedule as a teacher. I wondered when I could fit time in for training with everything I was doing.

Eventually, Cheryl let me know she was going to start a group of new volunteers in 2009 and I could be part of that. Even though I knew it would be a challenge to show up to all the trainings, I was committed to being able to serve the community as a firefighter and knew the knowledge and experience gained would be invaluable to living in Big Sur's wild landscape.

Trainings were held at the Fire station which by now had been completed. There was so much to become acquainted with that at first it seemed daunting. There was equipment to become familiar with, gear to learn to wear and put on quickly, radios to learn to operate, and a whole new way of communicating about everything.

The operational structure of the fire service, and for Mid-Coast, was also top down. I was meant to follow orders and operate within the hierarchy—not an easy task for a Stanford educated, type A personality.

We met every other Tuesday from 6:30pm until between 9 and 10. There was a mix of things to assimilate. Some evenings we would have two minutes to get into either Wildland Fire Gear or Structure/Vehicle Accident Gear and load into the engines for a simulation. We would extract people out of cars, feign medical emergencies, deploy hoses and spray water in different configurations, create fire line, run chain saws to remove brush, create burning house scenarios, and sometimes respond to real time incidents that came in while we were training.

Numerous weekends each year were devoted to extended trainings. We might spend a weekend learning to use the 'Jaws of Life,' and cut up a couple of old cars. HAZMAT training happened every other year, and keeping current as First Responders was similarly every other year.

Every week during my years on the brigade we were paged to respond to a variety of situations. Most commonly were the automobile accidents along Highway 1, or Palo Colorado Road, where the Fire Station was situated.

Though the time commitment was significant, the practical training and real-life experience was well worth it. When it came to responding to the Soberanes Fire as it came onto my property, the six years I spent on the brigade made all the difference. I was calm, prepared, and knew what to do to keep the fire from advancing.

Help Arrives

There were two downed oak trees on Long Ridge Road. Their bases had burned through far enough for the trees to fall. They had kept Jake and Cheryl from driving further up Long Ridge. If I could clear the trees out of the way, they would send help. I agreed to do so and they turned the chief wagon around to begin the descent down the road.

There were two large tow ropes in the 4Runner and with them I pulled the logs off to the side of the road. As I pulled the second tree out of the road, I looked down at my neighbor, Scott Rainer's property, and saw a few small fires burning very near the house. I drove down his driveway, parked, and began looking for a water spigot. I found one at the back of the house with 100 feet of hose coiled neatly beneath it. I turned the spigot on, allowing water to pass through the attached nozzle. There was still water pressure, and I assumed the water line was still intact.

I pulled the hose around the house and extinguished the small fires. Though none of them would likely have started the house on fire, there were six active fires within 20 feet of the house. I went to each one and put them out. Two were comprised of burning plastic. About 75 feet away from the house was a bunker built into the hillside. This was smoking. I walked over to it and saw within a smoldering pile of beams and a generator that had burned. I could not pull the hose over to this place to put the fire out completely, but it seemed far enough away from the house to pose no real threat.

There was also a fire creeping through the landscape below the house. After putting this out, I felt confident his house was safe.

I decided to drive further down Long Ridge to see what things looked like. I was stunned by the amount of burnt landscape. The landscape on either side of the road had burned completely. The ground was smoking with small fires still burning in tree stumps, and root balls. I stopped when I got to a downed telephone line. Two telephone poles had burned near their bases and toppled over, allowing the line to reach the ground. The downed line lay in the road, but still there was a narrow passageway for vehicles.

Soon Jake and Cheryl returned with my neighbor Scott Rainer. Rainer had been away on a trip with his wife and had booked a flight home when he heard there was a fire. He was a member of Mid-Coast Fire Brigade and made his way back to the canyon as quickly as possible not only to do what he could to keep his place safe, but to also help as a part of the Fire Brigade.

I took Rainer to his house. As I dropped him off, Wikar, Erin, and Jesse from Mid-Coast showed up to help. Wikar and Jesse, in Engine 7490, helped Rainer put out a fire in an oak that had fallen on an outbuilding and was still burning. Erin came with me to put out some small spot fires with the newly filled water trailer. We then walked to inspect the fire line I had created through the night and spotted a fire that had come around a rocky ridge. It was fire that spread from that little plume of smoke I had left.

Jesse and Wikar joined us in 7490 and we all looked over the situation.

The fire was burning in an area that had as its south boundary a lower road branching off the main part of Long Ridge Road. A fire road running through my property created the fire's north edge. I knew I could back down the fire road with my truck and trailer, pull hose from there, and extinguish the flames. Most of the land above the fire had burned making the proposition relatively safe. I presented that idea to Jesse, Wikar and Erin. Wikar and Erin were good with the plan, but Jesse was hesitant. Fighting fire from above is not always a safe practice, because fire tends to move quickly uphill when there is fuel to burn, but the land above the fire had already burned.

I asked if he would support the endeavor by providing some extra hose. He agreed. I started backing down the road—a narrow old fire road. With Wikar's help at an especially difficult sharp curve, we backed to a close and safe distance. We ran our hoses. I started the pump and Wikar started putting out flames. I joined Wikar, asked for the hose, and descended with it over the edge toward the active flames. Wikar followed me pulling hose through the burned brush. We made our way down through the burned area to the live flames hoping to extinguish them at a close distance. This made good use of our limited water supply. By the time the water ran out, the fire was mostly out. We shoveled through the area to secure our work, and then coiled up hose and returned up the hill.

Jesse had received orders to go check on other nearby properties and give a status report. To help assess the area, I drove down to the Jensen's property since Jesse had been advised not to do so in engine 7490. The road down to the Jensen's was too steep. I returned and reported to Jesse that Jensen's and Quinton's (another neighbor on Long Ridge) properties, looked good. I then drove toward Rainer's to see how I might help there. On the way to his place, I was met by a wall of fire coming across the north slope of Long Ridge and traveling east across my property. Flames were 5-8 feet high, and the fire had some momentum.

I turned around and drove back to alert Jesse, Wikar, and Erin. Jesse informed me CalFire strike teams would be arriving in minutes. I went back to the fire, and by the time I got there a CalFire captain and two engines had arrived.

From my vantage point on the road, I could see down the hill to an area where I stored a collection of useful things—17 telephone poles, steel girders, water line, redwood slabs, electrical wire, an old stove, a solar hot water heater, two 300TD Mercedes wagon parts cars, a trailer for my 32' sailboat, a 2500 gallon poly water tank, assorted lumber, 14 cords of cut and split firewood, and an old steel water tank I had converted into a storage building. The storage building was filled with an assortment of things but notably four tires and wheels for one of my trucks.

The fire was quickly making its way to that area. I felt the urge to drive down and defend the area. However, the flames were intense, and the fire was moving quickly. I realized there was nothing I could safely do, and little I could do to stop the fire from reaching it using only a shovel. I had to accept that I would probably lose everything there.

I looked on, standing next to the CalFire captain. He must have known what was going through my mind.

"I can't let you go down there, Theo," he said.

"I know," I replied, "I just can't help wanting to go down there and try to save things."

"Yep, I can imagine," he responded, "but we both know it's just not safe."

Then, something exploded. We looked at each other.

"Tires and wheels," I said, "I have an extra set for my truck in a storage container. I think one just blew up."

Another one blew, followed by another. The captain looked at me and nodded.

I watched as the captain and a couple of firefighters assessed the situation on the ridge. Another captain arrived and together they made the plan to back burn from Long Ridge Road to the north toward the fire progressing along the northern slope of my property.

Little did I know that this fire had been exacerbated by three CalFire fire fighters, using up their old ordinances (old equipment, in particular magnesium fire bullets they had been issued and wanted to use up so they could get new ones). This I learned almost two years later.

Aaron's Story:

It was the second Easter after the Soberanes Fire in 2018, when I had a conversation about the Soberanes Fire with Aaron, a friend in the Palo Colorado Community who lived across Rocky Creek from me. On Sunday morning, July 30th, 2016, the morning fire came up and across the north side of my property, Aaron had been close to my lower Northwest corner, on his brother's property which adjoined mine.

While he was there, a CalFire truck with three CalFire personnel showed up. They got out, readied their equipment, and began shooting ignited magnesium bullets up into the forest in the lower corner of my property. However the first bullet shot ricocheted off a tree and went under their truck, which they had to quickly move. After they had shot a number of bullets into the forest and created an intense blaze which gained momentum as it moved uphill, they turned to Aaron and asked if there were any houses above where they had started the fire.

His response was that there were a number of houses up the hill. They got into their truck and left. To this day I wonder why. Why did they do that?

Nothing more to do

I asked if I could help, and the chief suggested I get my truck with the full water trailer attached and head up by my house to sit tight. I liked that idea. I was definitely needing some time to rest. By this time, two engines had passed me moving very slowly up Long Ridge Road. I got in behind them to make my way up the road to my driveway.

As I drove up the road behind the engines, I found myself in the perfect position to watch the back-burning effort.

CalFire began the back burn (controlled fires set to eliminate the fuel in the path of a wildfire). Hot, high flames rose up along the side of the road. It was suddenly like being in a war zone. The heat, flames, and smoke were intensely hot even through the closed windows of my truck. Because the engines were taking up the entire road, I could not get past them to my driveway. There were two hand crews ahead of the engines and proceeding up the road. Men were raking sticks and debris into the area about to be burned. Five members of the hand crew were laying fire on the ground (starting fire with little drips of fuel poured from small hand carried containers). Others were shooting fire bullets down into the woods. As the fire was 'laid down,' the engines were driving along the road with one hose deployed and carried by a walking fire fighter. His job was to make sure the fire did not jump the road to the forest above. I wanted nothing more than to get my truck parked back up the hill near my house and get inside to rest. With all those fire-fighters working the fire, I felt I could be at ease.

Eventually space cleared and I drove up the steepest part of my driveway. As I parked the truck, I could see that fire had crossed the road and was burning at a lower corner of my upper orchard. I grabbed a shovel and ran that way. Four firefighters appeared walking up my driveway and onto the trail ahead of me to deal with the flames. I stopped and watched, breathing a sigh of relief as these men were responding to it and I did not need to. Even after the short run I had taken, my legs burned. It felt like my muscles were on fire. With the fever, when I ran or hiked up the steep slopes my legs burned so much I sometimes had to stop, crouch down, and wait for the pain to dissipate.

As I stood there, a fire engine drove up my driveway, parked, and deployed a hose toward the fire. I walked over to the driver/engineer monitoring water pressure at the back of the engine and asked, "Do you have enough water?"

He said, "I think I probably do."

"I have plenty of water and can fill you up if you need more water when you run out," I let him know.

"Wow, that would be great!" he responded.

"I'll get everything set up for you. The valve has a 2-inch female pipe thread. You'll need a fitting to interface with that."

"I think I have that," he replied.

So much for getting a rest! I set off to turn the necessary valves so the hydrant on the road would only need to be turned to start filling. When that was finished and all was ready, I returned to the driver and pointed out where the valve was on the road.

Through the rest of the day, back burning continued, while the fire crew made their way from house to house working by hand to make each one more defensible.

As night came on, I tried to rest. I still had not slept since spending the entire night stopping the fire from coming up the ridge.

The smoke was all around now, and it was hard to fall asleep. I was restless and decided to drive into town for some groceries. Who knew when I would be able to get out again? The thought of breathing fresh air was an incentive, and I felt like everything was safe with CalFire present on the ridge. As I left my property, I passed a fire engine parked in my lower driveway, which gave me more comfort. I drove to HWY 1 and checked in with the Highway Patrol Officer who was monitoring ingress. I asked if I could go out for some groceries and that I would be back soon. He gave me the OK.

I drove the 11 miles north to the Crossroads shopping area and parked at Safeway. Once I was making my way through the aisles, I began to feel like I'd made a mistake by coming in. My energy was very low. I was thankful I had a shopping cart to hold myself up. I got a few things and waited in the check-out line to pay. I was definitely in an altered state.

Items purchased, I got into my truck and started driving back. It seemed like it was taking a long time to drive back to the canyon, but what was worse; I did not recognize the landscape—the landscape I had driven past thousands of times. I wondered, 'Where am I?' I comforted myself thinking I could not miss Palo Colorado Road. Fortunately, I did not.

When I stopped to check in with the Highway Patrol Officer, he questioned me as to whether I had been out drinking. Though I told him I had not, he did not seem to believe me. I must have looked and sounded like I was completely out of it, which in fact, I was. He let me through and I wearily made my way home.

Dreams from beyond while everything burns

Back on the ridge, engines were posted along the road for the night. I thought I had nothing to worry about. I dropped into bed and fell asleep.

That night I had my first compelling dream during the fire. In my dream I became aware that I was looking at a powerful being that was fire itself—Agni (the Hindu fire god). His face was glossy, mask-like, and intensely hot. Though golden, the surface of his face was so hot it was shimmering like the rainbow you can see in black steel when it is extremely hot. The entire scene before me was fire, but within it was this beautiful, magnificently powerful being. He turned his gaze toward me and as he did, smiled reassuringly. Through his gaze, Agni communicated to me that he was doing his work—his passionate work.

Then he said, "This is my job, this is what I do." He went on, "In order to make way for the new, I have to get rid of the old." I nodded, affirming what he said.

He then looked directly at me and with a reassuring smile, finished, "Don't worry, but I am not done."

The dream was over. I awoke with those vivid images lively in my mind. Agni was doing his job!

Theo exhausted but keeping things safe

4

The Fire Saga Continues

I awakened Monday morning with the sun gently illuminating my bedroom through smoke-filled air. The smell of smoke brought me quickly into the reality of the moment. As I remembered what was actually going on, I did feel a sense of peace knowing firefighters were posted up all over our ridge, but I also knew that fires were unpredictable and often uncontrollable. I got up, pulled on some clothes, and headed off to the kitchen. Feeling better than I had, but still feverish, I sat sipping warm water and wondering about what was next.

Minutes later, Cheryl and Jake Goetz knocked on my door. Cheryl let me know that she needed me to help as a member of the Fire Brigade and did not want me worrying about my animals. She had already been on the phone with the SPCA. With little persuasion, she convinced me to let the SPCA take my horses to safety at their facility on Highway 68. None of us knew what would happen next, and it was clearly a good idea. The horses had seemed pretty mellow through the couple of days of fire. They stood through it huddled together in their paddock watching the activity. They never got wild or even whinnied, but better to have them safe and far away from the fire.

Cheryl made final arrangements from my phone line which was miraculously still working. After she had arranged a time for me to meet them at the bottom of Long Ridge, she and Jake left.

Unbeknownst to me, Jake and Cheryl had also contacted a capable neighbor, Travis Trapkus, and recruited him to show up and help me out. He soon arrived and we began the process of loading the horses into the horse trailer. After a few minor difficulties, we were soon down the two miles of steep dirt road to Palo Colorado Road. Within minutes the SPCA arrived. We transferred the trailer from my truck to theirs and off they went, down the canyon to safely keep my horses for the duration of the fire. I was thankful there was one less thing to worry about, and that Cheryl and Jake had made the calls to arrange the transport.

More fire crews came and continued back burning around houses up on our ridge, starting with the Quintons' and then the Jensens'. I am not sure when they finally got to mine, but by the time they did, I was asleep. When I went out to inspect, I was shocked to find they had burned the perimeter of my garden and orchard—an 8' high deer fence. They burned two of my irrigation timers, which had open valves, allowing the water in my irrigation tanks to drain completely. They burned lavender plants and a number of fruit trees in pots. I could not understand this. Who trained these guys? The crews doing the back burning were inmate crews. From what I understood later, one crew was well trained and did a careful, respectful job of back burning, and the other crew did not. I unfortunately got the crew that did not. What the crew on my property did was totally unnecessary, but trying to stay positive, I focused on the fact that my house and all auxiliary structures remained intact.

By the end of the next day, it appeared from the level left in my tanks that between what the firefighters used and what had drained out of my tanks, I had 8000 gallons less water than I did when the fire began. I thought it a wonderful thing that my rain catch water system provided enough water for the fire fighters to effectively do their job, but was dismayed that I had lost so much when my irrigation timers were burned.

Monday night I slept but tossed and turned often. The smell of smoke filled my house, and it was hot. The fever continued, my temperature never dropping below 101.

Tuesday morning, the mountaintop was engulfed in smoke. It was difficult to breathe. I spent most of my time laying in the house, consumed by the fever that now reached 103.2. By mid-morning the phone line went out. I had managed to keep in contact with a few people, including Patricia Bercovich who posted updates about my status on Facebook. I had also talked with my son Chris, who offered to come help. Feeling horrible that morning, I wished he was already with me.

I did go out to drive through the neighborhood and make sure things were OK, but I was exhausted from the effort.

Mid-morning, I was down by my lower gate, extinguishing fire burning in a post, when Travis Trapkus appeared on his quad with a Monterey Herald photographer sitting behind him.

Travis lived down in a lower part of Palo Colorado Canyon in an area that was not currently threatened by the fire. He was low enough in elevation that the moisture content in the vegetation was high enough to make it more difficult to burn. Travis had helped me out with some of my concrete projects, bringing his talents into play to turn what would have been a concrete job into a work of art. He was a master artisan with concrete, had served on Mid-Coast Fire Brigade for a couple years, and was a solid community member.

"How's it going Theo?" Travis asked.

"Dandy," I replied, taking advantage of the moment to lean against my shovel and take a break. David Royal busily snapped pictures and Travis joked about my becoming famous. We talked a little more and Travis asked me a question that he continued to ask every time he saw me during the fire.

"Theo, do you want to save your home?" he would ask.

Of course, my response was, "Yes."

To which he would reply, "Then stay home."

After a little more chatting with David Royal asking me some questions about the fire, Travis turned and drove back down Long Ridge.

A day later it turned out that one of the pictures David snapped appeared in newspapers and online articles, including the LA Times, so for at least a moment I had my claim to fame as a bedraggled property owner defending his home from fire.

Later, after returning home, the power shut off in the house. The electrical system for the property was an off the grid solar system with a back-up generator. I went outside to start the generator. The smoke was so thick that the solar panels were not charging the batteries. I came back in to rest for a half hour while the batteries recharged. A half hour

Theo makes it into the news

slipped by. I was aware of this and kept thinking that I should get up to turn the generator off, but it seemed my body would not respond. Finally, with Herculean resolve, I pushed myself up to sitting, stood, and, wobbling like a drunk, made my way to the door. I began walking toward the generator shed grabbing what I could for support. I was weaving and could barely keep upright. I walked twenty paces and wondered where I was going. It took me a bit to remember, and I then wondered what was going on with me. I had never experienced anything like this before. I kept walking and finally arrived at the generator. I could see the button I needed to push to turn the generator off but felt frozen—stalled and unable to move. Determined and with great focus, I leaned down and pointed my finger toward the button. With great effort, I got my finger to the button and pushed. The engine died, I closed the doors and turned.

The next thing I became aware of was something caressing my face. Before I knew what it was, I thought, "Wow, this feels good." It turned out it was my dog Azu's moist tongue. I came to consciousness enough to realize I was lying face down on the ground. I couldn't get up. I began crawling over to a concrete pad a little way ahead of me and slowly pulled myself upright, leaning against a post as I did so. The front door of the house was about 60 feet away. With Azu beside me, I pulled myself from one object to another and then stumbled the last 20 feet to the door. I made it through and went in and sat down.

What just happened? There was a scrape along my forearm from my elbow to my wrist. The blood had already dried. I must have been unconscious for at least a half hour. Was the oxygen content in the air too low, or carbon monoxide levels too high to support consciousness?

I needed to get off the hill, and to do so as soon as possible. I needed to get some fresh air—house or no house.

Monday afternoon, when the phones were working, Chris said he would come from Sacramento to help. Maybe he had arrived. With no way to communicate with anyone, there was one more reason to go to town. It was certain it would be helpful to have him there with me.

Driving down Long Ridge, there were two places where the fire was actively burning the south side of the ridge and up to the road I was driving on. I stopped and watched. The fire would make an accelerated run uphill to the road before slowly backing down to the bottom. It would then repeat the process.

Two fire fighters were posted along the bottom portion of Long Ridge Road. One engine was monitoring a small active burn, while another engine was posted along the road watching for spot fires. No firefighters were at the top of the ridge where the remaining houses stood. Fortunately, though, some neighbors had returned.

Arriving at the bottom of the dirt road, I took the truck out of 4-wheel drive, and started down Palo Colorado Road. I checked to make sure my lifesaving face licker, Azu, was there behind me, slowly driving down Palo Colorado Road to Highway 1. I checked in with the Highway Patrol officer monitoring ingress and made sure I could get back in. The air quality near the ocean, and at this much lower elevation, was pure and agreeable. Breathing in that sweet air was already doing wonders for my body. Coming back to life, my thoughts became clear, and I started to feel good, convinced that it must have been the poor air quality that caused me to lose consciousness.

I headed north beneath the smoke-filled sky, where dim light prevailed. At my property which commands amazing long-distance views, I had barely been able to see 100 feet. Coming out of that smoke-encased landscape, and driving along the fire's perimeter, one could see the active fires on the hills and in the valleys along the coast. Fire equipment was everywhere along Highway 1.

Once in cell phone range, I called Chris and left him a message and a text, letting him know I needed his help and to please come. I also detailed what had happened earlier in the day with my loss of consciousness, in the hope he would realize the severity of the situation.

Next, I sent a text to my friend Kierstyn Berlin, who had a small place in Carmel. "Dinner?" I inquired. As my body regained some vitality, I found myself very hungry. She responded affirmatively.

We met at Le Bicyclette, a local favorite. As I entered, and the beautiful smells of well-prepared food filled my senses, I felt ravenous. The receptionist let us know there was a 20-minute wait, but as luck would have it, two of my former students, Emily and Kate Burton, were working that night as servers. They saw me. They knew I was a firefighter, knew the fire was burning the area where I lived, and had been following my saga through Patricia's Facebook postings. The two of them surrounded me in a hug, and had us seated immediately. They were quite demonstrative in their concerned affection and inquiring questions. As this all unfolded, the other people at the restaurant turned their attention toward our table. I was quite a sight I am sure—I smelled, and I am sure looked like I had just come out of the fire. I had not bathed before coming in, and later when I looked at myself in the mirror in the bathroom, saw that my face was smudged with black. And my hands, even after having been washed, were etched with black.

As I told Kate and Emily about what had been going on, everyone around us listened attentively. I fielded a few questions from my captive audience before everyone seemed satisfied and returned to their meals.

Emily and Kate were relieved to hear my house was still standing. Much of the redwood I had used in my house was milled from their family's land in Aptos, California.

Soup, salad, and pizza made for more food than I could stomach. Since the fever began, I had barely eaten, and after every 2-3 bites I simply had to wait before continuing. Away from the fire, in good company, having a delicious meal, I found myself feeling that it was still good to be alive.

As we sat at Le Bicyclette, Chris texted. He was on his way from Fair Oaks but would not arrive until midnight. We finished dinner and headed outside. The fresh, moist, cool air of Carmel was like heaven-sent prana. Kierstyn offered to make up a bed for me. I was soon lying down in a place far from the fire, with a cool breeze blowing in through open windows. I was dozing off when Chris arrived. The three of us visited a little before we all went to sleep.

Chris and I left Carmel around 8 am after a small breakfast. We drove straight to the fire station and checked in with Chief Goetz at Mid-Coast Fire Brigade to see what we could do to help. It was a little strange to arrive in the canyon and not go check on my place, but I was still on -Mid-Coast Fire Brigade and felt like I needed to do my part to help the community. Certainly, the neighbors who had returned to Long Ridge were keeping things safe.

After a brief check-in, we were assigned to keep three 1500-gallon portable water tubs filled, and to fill all engines and water tenders that came to our filling station, along Palo Colorado Road, next to Rocky Creek. There was a pump with an intake line in the creek that could either fill an engine directly from the creek or fill any of the three filled water tubs. Chris and I went over the operation of the system together. If he could run things by himself, it would give me a chance to try to get some rest in the truck.

Through the day we repeatedly filled two water tenders that were making continual runs to the fire line. Occasionally a fire engine would arrive. We had hoses in place before the trucks had stopped and had pumps running within seconds of attaching hoses to their inlets. As in most firefighting proceedings, time is of the essence.

We did, however, have time to get to know the drivers of the water tenders and engines as the tanks filled. Rob, one water tender driver, would give us updates as we filled his truck. The fire had intensified through the day. That became ever clearer as we were soon working non-stop to keep the water flowing.

Sometime in the afternoon, Brent Bispo came with Mid-Coast Fire Brigade's engine 7411. He spoke quietly with Chris and me, letting us know they had been battling the blaze and had kept it from burning the Goetz house and other homes along the ridge. He said we were doing a great job filling the trucks, as there had never been a moment when they did not have water. He was exhausted and relieved to have a moment to take a break. He said things had mellowed out a bit, so now would be the time to go to the station and get more fuel to make sure the pumps could keep running.

Informal meeting in a patch of green at Theo's property

Aerial view of burned area, and fire progressing to the south and east. Theo's property marked.

5

A Death and Destruction

I jumped in my truck as soon as Brent left and made my way to the Midcoast Fire Station, pulling into the driveway and close to the shed where containers of gas were stored. There was an ambulance parked there. As I got out of my truck one of the ambulance drivers approached me. He was a friend.

He took a good look at me. I must have been a sight. I had black Carhartts on showing the wear and tear of the last couple of days. My hair was a mess and I had a Mid-Coast Fire Brigade t-shirt on with circles of wet at my armpits. He asked how I was doing.

"I'm doing great," I returned, with an overtone of sarcasm.

"I can see that," he said. "Tell me what's going on. You look like you have been through hell and back."

"Thanks for that," I said, "I feel that way and sort of like I am still in hell. I have had a fever since the fire started, have hardly slept, and have a raging headache!"

He followed me over to the shed and when he saw I was grabbing containers of fuel he took them from me. I lowered the tail gate of the truck, and he loaded the two five-gallon fuel tanks in. I closed the tailgate and walked to the driver's side door to get in and drive back. As I did, a parade of CalFire trucks came down from Green Ridge Road, the road directly across from the fire station.

A bulldozer operator had rolled his bulldozer cutting a fire line and sadly died. They were bringing the dead bulldozer driver off the mountain. Strangely, he had died close to where another bulldozer operator had rolled his dozer and died in the Palo Colorado Fire of 2010.

Palo Colorado and the road up Green Ridge were now choked with trucks. As the trucks parked, a large group gathered, completely blocking the road in all directions. There was no exiting the fire station to return with the five-gallon jugs of gas.

My friend asked if I wanted to sit in the ambulance for a bit and I agreed to do so. Knowing I was not doing well, he asked if I wanted some oxygen. I agreed, reclining in the seat of the ambulance as he fit a mask over my face. I started to breath the straight O2 and immediately felt a cool wash of relief cascade through my overheated body. My headache almost disappeared.

I watched through the ambulance windows as the men below performed a ceremony in offloading the dozer driver out of the sheriff's truck and into the coroner's van. It was military style with a folded American flag and four men on either side of the body. I had plenty of time to breathe deeply and regain some ease in my body. As the transfer of the body of the bulldozer driver neared completion, I knew it would soon be time to detach from that sweet flow of oxygen.

When the coroner's van pulled away, I handed the mask back.

"Feeling better?" my friend asked.

"Way better," I said.

"You look way better," he replied.

Thankful, I bid farewell, and slowly drove back to the pumping site. Chris was sitting on top of a water tender watching to see when the tank was full. I re-joined him.

It was great to work side by side with my son. We had worked a lot together through the years. This day we had fallen into an effortless flow of sign language to get our job done. I couldn't tell him enough how glad I was he was there.

The day sped by. It was difficult to tell time, as the sky remained a sullen grey. By eight, the sky was darkening. I made my way back to the fire station for updates. I was instructed to get enough fuel to keep the pumps running through the night, and to get some provisions. There were platters of fresh food spread out on a table in one of the engine bays. I dished up a pan of food for us, grabbed another five-gallon jug of fuel, and returned to the pumping station.

I ate a little on the way back. I was sort of hungry, but everything I ate seemed to make my stomach and esophagus burn. I wondered what I was fueling inside my overly hot insides. Mostly I did not feel like eating and even drinking was a bit of a challenge. I could drink a few gulps and feel nauseous, or eat a few bites and have my stomach burn. It was not a good recipe.

Just as I was getting back and walking over to share food with Chris, Cheryl returned and asked how we were doing. I told her we were good and that I had food from the station. She asked if we had sleeping gear. I let her know we did not. She instructed me to drive up the hill to my house, get sleeping gear and come back down. She said to leave Chris so he could fill more trucks if necessary.

By now it was dark. Driving up Long Ridge toward home, there were still fires on either side of the road. Reaching a good vantage point close to the top, I slowed to fully take in the scene. From there it was clear most of the fires in the near vicinity were out, but there were still things burning in places throughout the landscape. Once home, I parked the truck in an advantageous spot to load up. As soon as I got out of the truck, I smelled gas and realized I had forgotten to leave the five-gallon jug of fuel at the water pumping station. I opened the tailgate and found the jug on its side. Gas had leaked out into the bed of the truck. *Fabulous*, I thought to myself. Nothing better than spilling gas into the back of your truck while driving along a road with flames on either side, and then the thought came that we were likely going to be sleeping in there. I took the jug out and hopped up into the bed, throwing out a couple of other things. I threw some dirt onto the gas, then ran and got some detergent and a brush and hose and scrubbed the gas-covered areas with soap and a little water. I then washed out the whole back end as it was covered in dust and dirt. All the dozer and engine traffic on Long Ridge had turned the top 6–8 inches of the road into a downy fluff. It made for a very soft and dusty road.

Leaving the truck to dry out I went to get sleeping bags and pads. I loaded them into the cab and thought it best to get some fresh clothes.

Stepping out of the house with a bag of fresh clothing and a towel, my neighbor Zack pulled up on his quad. "Theo, we've got fire right over there. Over in your orchard." He looked back over his shoulder, and I followed his gaze. Sure enough, there were 3-5-foot flames just above the road, with a small trail between it and more dry tinder. The fire was burning just below an upper orchard. If the fire continued another 100 feet up the hill, through the orchard, my neighbor's house would be on fire and the entire ridge threatened again.

Over the previous two years, after clearing brush and dead trees from about an acre of land on that north sloping hillside. I terraced the cleared area and planted an orchard, which now had a mix of 60 fruit bearing trees. That clearing turned out to be a home-saving endeavor—the fire had little fuel to proceed up the hillside, but there was enough fuel on either side for the fire to make a run.

The active fire was just at the edge of the orchard and close to the remnant piles of brush and debris. Shovels in hand, Zach and I pushed and shoveled the burning material over the bank and into the road. We stood back and watched as the material burned itself out. We then created a four-foot line above where the fire had been burning and pushed the remaining unburned fuel over the edge and into the road. In about 30 minutes, we had the fire extinguished.

Fires change relationships. When you are faced with something as potentially disastrous as a fire, conflicts are replaced by collaboration. On Long Ridge, once neighbors came back, there was an overwhelming feeling that we were in it together. We were all watching out for each other. Zach and I had not always shared the same views, but after we faced the Soberanes fire shoulder to shoulder, a bond was formed that has since not been broken. Our differences are acknowledged and appreciated, and any animosity has been replaced by goodwill. When he needs help with something I show up and help. And he does the same for me.

Once we were certain the fire was out, we walked back to where my truck was parked. We caught up on the status of the fire and then parted ways, Zach returning to his patrolling of Long Ridge while I continued to load gear into my truck. As I did, an intense orange glow began to grow to the south. The fire was coming around the ridge from the south!

Agni's words came back to me, "Don't worry, but I'm not done."

CalFire had spent a good amount of time and manpower clearing and back burning around people's houses. I knew there was burned material and cleared areas around almost every house up on the ridge, but CalFire had focused their efforts on homes they thought could be saved. There was a lower road that branched off Long Ridge Road. There were two residences along this road. CalFire did not deem either one defensible and, therefore, had not back burned around them. Both residences were below my property.

The flames intensified, rising 80 feet into the air with periodic fire bursts in the thick smoke. Red hot embers sailed through the air, creating the possibility for spot fires in any unburnt area. I went to a steep place in the landscape, which looked directly down toward one of the residences. An outbuilding was on fire, and nothing could be done to save it.

Unfortunately, with the huge flames, I was worrying. It looked to me like anything might happen with a fire like that. How could I return to the pump station for the night? I felt the need to stay on my property.

I remembered what Travis had said to me, "Theo, if you want to save your home, stay home!"

If the fire continued undeterred, the other buildings of both residences were likely to burn and there would be more intense episodes. I had just put one small spot fire out on one side of my property and now fire was actively coming around from the south.

I waited and watched until whatever had been burning so intensely, mostly burned out.

I headed for my truck. I checked to make sure Azu was loaded in and that I had tied the gas tank securely in the back. I started down the hill with the intention of getting Chris and heading back up. I hoped to be able to tell someone of my intentions on the way. I wanted to get permission to leave our pumping station.

Nearing the bottom of Long Ridge, I drove by Jake and Cheryl's. Jake, assistant to the chief, was standing outside on the road. I stopped the truck.

"Jake, I am so glad to see you. We have fire up on the mountain again. It's coming around the south side and it's burning toward the houses on the lower road. I want to go back up with Chris and make sure it doesn't catch our little island of safety on fire. Can I leave the pumping station and go back up?"

"Can you guys handle the fire up at your place?" Jake asked.

"I think so," I responded.

"OK," Jake said, "Then head back up, keep safe, and come back down if you need help. I can always pull someone off from what we are doing to help you out."

"Thanks, Jake."

I started my truck and drove down to where Chris was standing with his arms crossed looking as if he were going to freeze. As night fell, the lower canyon had become cold. I had left him pumping cold water out of Rocky Creek and his sweatshirt was in my truck. I came back about an hour later than I had hoped. Strange, how in the midst of an intense fire, down in the creek bed it could be so cool and so moist that you end up cold.

Chris was glad to see me. As Chris was getting into the truck, Travis showed up.

He looked at me and asked, "Do you want to save your house?"

"Of course," I answered affirmatively.

"Then stay with your house, Theo. If you're there it is not going to burn. You've got the most defensible house on the ridge. Stay with your house!"

"We're heading back up there now" I said.

Travis sighed, "Good, I'll be up to check on you."

Chris and I returned up the mountain. When we got to the top and parked the truck, it was clear that one of the houses was burning. The flames were shooting 100 feet straight into the air. The glowing hot particulate being carried up into the smoke column looked like an angry genie let out of a bottle. We watched, shocked into the reality of what was happening.

We pulled our attention away from the fire to focus on the tasks at hand. In between our inevitable glances toward the glow, we kept doing things to minimize the chance of a fire starting within the unburned area around our structures. We moved brush away from the edge of the mountain and closest to the active fire, repositioned hoses, walked over the entire property, and drove through the neighborhood checking for fires.

I still had a fever. Any attempt to hike up a hill, had me stopping every 10-15 steps because my thighs and calves would be burning as if I were running a marathon. It was so painful, driving around to the places we needed to go was a better option than walking.

It was very smoky, making breathing difficult. There was no fresh wisp of air to come sallying by to relieve our burning throats and lungs. Without goggles our eyes burned. Chris, who could barely see, put on the one good pair available. Chris asked for a respirator. It was a good idea. We both put one on and began clearing away already cut manzanita from under the tree house built with my stepdaughter, Julia. As we moved brush, there was a tremendous explosion. A propane tank had blown up.

From the first house, the fire continued west along the unburned brush until it consumed the next house. This sent huge flames into the air and two propane tanks began to bleve[1], sometimes in unison. Chris had never seen or heard anything like what we were witnessing. He wavered between awe and fear. We continued to create a larger safety zone around buildings until we thought we had done all we could. We even cut down a beloved Bay tree close to the house.

When we went into the house the carbon monoxide alarm was going off. We stopped it and took it outside and it began to beep again. Not a good sign.

It was 4 am. We had done all we could and had been working since we arrived in the canyon in the early morning hours the day before. Both of us, though wide-awake, thought it would be a good idea to take a break. Inside the house things seemed unreal, with smoke as thick as it was outside. I thought cleaning myself up might help make me feel a little better, so I readied a bath. Afterwards, as I pulled on fresh clothes, I urged Chris to also take a bath, but he was on high alert, walking through the house and looking out the second-floor windows.

"Dad, the barn is on fire! There is fire on the barn!" Chris suddenly called to me.

From the house we could see bright yellow flames about a foot high coming from a corner of the barn. It was about 400 feet away and downhill. We pulled on shoes and ran down the hill. The fire was burning in the copper gutters. An ember must have blown up from below and landed in the gutter, catching the leaves I had not cleaned out, on fire. It was the one roof I had not cleaned on Saturday—the day the fire hit the ridge. We quickly got a ladder and a couple of masonry tools that would act as scoops. Chris got onto the roof and began dealing with the fire.

I made my way as fast as I could back up the hill to get 50' of garden hose. I was frustrated that I had not already brought one to the barn. I returned with it in the 4Runner.

Chris pulled the leaves not on fire out of the gutter. Then he spread the burning leaves out inside the gutter. I ran with the now connected hose to the barn. Chris hopped down off the roof, grabbed the end of the hose, and began to climb the ladder. Something weird happened while

---

1 A boiling liquid expanding vapor explosion caused by the rupture of a vessel containing a pressurized liquid that has reached a temperature above its boiling point.

stepping onto the roof. I heard a loud crash as Chris fell backwards onto the deck. He immediately sprang back up, saying he was OK, repositioned the ladder, and made his way up again to spray the gutter with water, making sure everything was out. Luckily the barn was constructed of fire-resistant materials—a steel roof, with steel and stucco siding. If built with more conventional materials it would surely have burned.

We hung out for a little while near the barn. From its deck, we could see the active fire a hundred yards below us. Between us and the active fire almost everything had already burned, and the house directly below was now just a pile of glowing embers. The threat seemed over. We drove back to the house. On the way I checked in with Chris to make sure he was OK. He had hit his head as he fell, and it was bleeding slightly. I looked more closely at it with a headlamp and saw that it was a scrape more than a cut. It had clearly been a pretty good concussion to the head.

As we got out of the 4Runner, we noticed that it was brightening to the east, but it was not a pre-dawn light. Fire had again started in fuel below the new orchard. This time the flames were larger, and the area burning was bigger. I tried to see if the fire hose we had already laid would reach the fire, but it did not. We grabbed two shovels and walked over to the fire.

Chris had never worked a fire so closely. I explained what we would do as we approached. The plan was to work down to bare mineral soil above the fire, shoveling the soil onto the fire. We would also shovel the burning fuel back into the already burned area. We quickly dug the line and then began shoveling the burning fuel into the already burned area. Some we pushed over the edge of the embankment and into the road. Having done what we could, we sat down and watched to make sure it would burn out. We worked the area a little more with shovels and were soon satisfied with our efforts.

The sky began to brighten—this time with the morning light. It was 5am. We walked over the unburnt island of green and around all the buildings. It seemed to both of us that the place was safe, and the spot fires were out. We had been up all night. Again, our efforts had been crucial to saving the structures and perhaps the rest of the ridge from burning. We hopped in the 4Runner and made another trip through the neighborhood, driving all the way down to the Jensens' and all the way down the lower road to one of the burned homes. Jensens' and everything in between my property and theirs was quiet with a few large pieces of wood smoldering. On the lower road both residences had burned completely, including a shed built illegally on my property. At least there would no longer be an issue with what to do about that.

In 2009, while building my barn, someone turned me in for building it without permits. As a result, I was required to do many things to bring all I had done into compliance with Monterey County codes. One of those necessary tasks was to have my property surveyed, which proved to be a great thing, clarifying where the property lines ran through the landscape.

It turned out that there was a shed built on my property, but claimed by the neighboring property. I offered to give the neighbor the land that the shed was on for free, if he would just take care of the paperwork. I thought it a generous offer. I received an email back letting me know if I touched anything of his, even if it was on my property, he would sue me. I thought the response odd, and had simply let the whole thing alone, hoping it would somehow be resolved.

The fire took care of the problem completely.

Dreamscape

6

Another Vivid Dream

Feeling confident all was well on the ridge, Chris and I went into the cold room at 5:30am. Earlier, we had pulled two futon mattresses into the room, and Azu was already set up in there on her bed. We minimally opened up one of the scuba tanks and lay down. Within minutes the air in the room became cool and clean as the compressed air filled the room. While the tank slowly emptied, the decompressing air cooled the tank to the point that ice began to form on it. I felt better than I had felt for days in that cool room. We talked for about 30 minutes, going over the events of the day. Chris had never experienced anything like this, and he had been amazed at the intensity of the fire—especially when the houses below us were burning, and the propane tanks had bleved and then exploded.

Before long, fatigue won out and overcame our ability to talk. We were soon asleep.

The Dream

I stood on a mountain ridge. It was beautiful and though I did not see any people, I knew I was in the mountainous terrain I had grown to love—Big Sur. I felt full of love and at peace. It was a beautiful scene looking out over the landscape, through the forest to the ocean, and I felt one with it all. I became aware that Mother Earth was before me. She was the earth, but I could only make out the back of her head and her dark glistening hair. Her long hair was flowing into a fold of the earth, and the surface of the earth was rolling into this fold. It looked like a breaking wave, made up of earth instead of water. What I watched was changing rapidly, giving me an uncomfortable feeling. I wanted things to be still and changeless, in part because it was so beautiful just the way it was. I also did not like seeing Mother Earth's beautiful hair getting turned into earth.

I asked Mother Earth if she would stop for a moment, because it was beautiful just the way it was. She stopped making the earth move and change, and gave her attention to me.

She said, "I will stop for a moment, for you, but I cannot stop. Change is what keeps all of this beautiful."

A moment later her beautiful dark hair was again being pulled into the earth. This time I was at peace while I witnessed the never-ending change. I awoke.

I lay still, taking in the feeling of the dream. Chris soon stirred and looked over at me.

"Good morning," we greeted each other.

Our sleep had not lasted long. It was 7:30am and after talking a little, we felt like it was time to get up.

Chris and I donned clothing and boots and opened the door. It was warm outside and still very smoky. We went into the house, had some water and a little food, and talked up a plan for the day.

7

Mid-Coast Fire Brigade

As we surveyed the smoky world in the morning light, everything seemed quiet and peaceful. The night's intensity had passed, and though we had only slept a short time, we were refreshed.

We had no way of communicating with anyone. The phone line was out, and there was no cell reception up on the mountain, or for that matter, in most of Big Sur.

Chris was concerned that people would be worried about him—especially his fiancé, Linsey. He wanted to go into town to let everyone know everything was OK.

We decided to each drive one of my trucks down the hill, check in with Mid-Coast to see if we were needed, and then get Chris into town. I would stay in the canyon and close to home.

I had Chris drive the 1989 Ford for the town run. We compiled a list of things for him to get while he was there—mostly water lines and supplies for people on Long Ridge whose water systems had burned. As we drove down the hill, all seemed well. We did not see any active fires until we were over halfway down Long Ridge and these were inconsequential fires still burning the south flanks of the ridge.

We checked in at the fire station and were told we were not expected to do anything that day and could have the day off. Chris was ready to go to town, craving some fresh air. We both drove trucks down to Highway 1 so I could negotiate with the Highway patrolman for Chris to have permission to get back into the canyon after his run into town. We parked near the Highway patrolman and went over to talk with him. John Hains, a canyon resident, and his son Conrad were there trying to get up the canyon to their place. John wanted to get his water system up and running. Their water system provided water to the Fire Station's water tanks.

"Theo, can you take me up to my place?" John asked.

"John, I wish I could, but I can't. I can go back and ask if it's OK for you to come in, and come back and let you know," I responded.

I could not imagine a more important water system to get running again, but I did not have the ability to get John in just because I was a fire fighter. The Highway Patrolman had made it clear he would not let John in. Many people were requesting special permission to get into the canyon, and he was clearly not going to budge. John asked me, within earshot of the patrolman, if I would at least take the new battery he had with him, up to his property. Before I could respond he was coming to the side of my truck with the battery. I opened the door to put it on the floor.

"I'll meet you at the Grimes Ranch Gate, by the bridge," John quietly whispered to me. The Grimes Ranch gate he was referring to was on the 'inside' of Palo Colorado Road. He and his son Conrad went back to their car, drove to Highway 1, then turned to head south.

While all this went on, another friend and neighbor, Rurik Draper, was trying to get in. We looked at each other. I pointed to Chris and then toward the Grimes Ranch gate. Rurick gave me a thumbs up and went back to his truck.

Earlier during the fire, while I still had a working telephone line, Rurik and his wife Allie called to ask if their place had burned. They lived about a mile from me, down Long Ridge Road toward Palo Colorado Road. From what I witnessed Saturday night, when the fire was extremely active below me, I had assumed their house burned. It just didn't seem possible that it had made it through. But I was wrong, and on Sunday, when I went into town, I drove by their place to see if it was still there.

It was.

I called them as soon as I had cell reception in town, but felt horrible for what I may have put them through by telling them earlier their house had possibly burned.

I explained to Chris what would happen next. These neighbors were going to run across and through Grimes Ranch to the gate, a little way up Palo Colorado Road, and beyond view of the Highway Patrol officer, who I had not had the chance to speak with. I would take John and Conrad to the Fire Station to see Cheryl, and Chris would take Rurick and Allie to their house up Long Ridge.

I was conflicted. I was bringing people into the canyon without our fire chief's permission. It was not acceptable fire fighter behavior. Though our chief had always encouraged us to think on our own, I knew this would not meet with her approval, but as a community member, I just couldn't say no.

We turned our trucks around and drove back up the road. I pulled into the small turn out, hidden from the Highway patrolman. Chris was right behind me. John and Conrad piled in with me, and Rurick, with Allie, piled in with Chris. Up the canyon we drove. I pulled in at the fire station so John could talk to Cheryl. Chris drove by me with Rurick and Allie.

After about 20 minutes in the station, John reappeared and hopped in. Cheryl had given John permission to get his pump going. I drove John up to their generator room with the new battery. He explained the system to me, showing me everything, in case something needed to be done in his absence. I left the two of them to sort things out and headed back to the station. Cheryl had said she wanted to talk to me, so I stopped at the station on my return.

As I suspected, I had upset the chief and I got an ear full. Bringing John in had not been OK and she was already upset by other things, including that Chris had come in with me the day before, that we left our pumping station and headed back up to Long Ridge that night, and that I drove too fast on the road trying to get back to my place with Chris when the fire was threatening our island of green. I was not following her protocol and she was not happy about that.

The way I looked at it, John needed to come deal with his water system so the tanks at the station were full. I figured bringing him in saved everyone time. Though it may have been difficult for people to believe it, I was really ill. I needed Chris there to watch out for me if I was going to stay at my place to see that things did not burn, and to help out the Fire Brigade when I could. 'When I could,' is not really an acceptable response for a firefighter. Ideally, I would be available and in service to Mid-Coast Fire Brigade rather than worrying about my own place. But the night before, if Chris and I had stayed down the hill pumping water, my barn and orchard would certainly have burned. And if either burned, the fire would have gone into the unburned area of the ridge and the whole ridge could have gone up in flames. If Chris had not seen the barn on fire when he did, it may well have burned, even though it was built out of fire-resistant materials. I was quite certain if Chris and I had not been up on Long Ridge that night, things would have been quite different.

Driving too fast—I never really drove too fast, just fast for someone who did not know the canyon. I was actually driving slow by my standards. On Tuesday night, when Chris and I were trying to get back up to my property to make sure the newly active fire did not burn the ridge, the Incident Commander for the night made a comment to me that I had better slow down. I needed to get back up the hill. I was too sick and too tired to get into it with him, and couldn't have cared less what he thought. I blew him off by not responding and simply driving away. He reported my rude behavior to Cheryl. At that point, I did not care.

I was willing to take responsibility for my choices and actions. I thought I chose well and wisely. But I also understood Cheryl's perspective. The fire service, like the military is a top-down organization. If you do things on your own and do not follow the chain of command, things will not go smoothly. I did not do well with a top-down hierarchy. I was far too independent—a free thinker. I made my own decisions and though they were, most of the time, what Cheryl would have supported, she probably felt like she didn't know what I would do next. I imagine she saw me as a liability, and I could understand that.

After she finished sharing her thoughts, I was angry. I had done what I thought was best, I still had a fever, I had hardly slept the last 5 nights, and I felt unappreciated for the effort Chris and I had made pumping water. I left the station, feeling I was done serving on the Fire Brigade—too much conflict and no peace. I had been thinking about leaving the Brigade for a while. I knew I did not quite fit the firefighter mold, and it took too much time away from things important to me. But the thing that had the most impact on my decision was that over the course of six years, showing up on disasters had taken a toll on my psyche. I had developed a serious case of PTSD.

Post Traumatic Stress Syndrome is serious. Over the time I served on the fire brigade, I witnessed myself go from being happy go lucky and very positive, to not feeling like life was worth living. Mornings were the worst. I just did not want to get up.

As a fire fighter, you do not just show up to put out fires. You show up for vehicle accidents, medical emergencies, accidents in the wilderness, drownings, and situations involving domestic violence.

You never get paged to come enjoy a beautiful meal, or a walk on the beach at sunset.

I showed up for 1-3 calls per week and with that steady diet of disasters, I developed a mind-set of *What messed up thing is going to happen next?*

I had seen horrible accidents with people severely injured and sometimes unconscious, helped extract injured people out of the wilderness, lifted dead people into body bags, helped load ill people into ambulances, helped load injured people into helicopters, watched people taken away by sheriffs in handcuffs, and shown up to structure fires, car fires and wild fires.

It became clear to me that we live in a world of constant change and that extreme changes can happen very quickly. One might say, "Wow, that sounds like enlightenment. Just surrender to the change."

The problem for me was that I began living with the expectation that something bad was going to happen, waking up every morning with this gnawing feeling inside like there was no point in getting up, because there was nothing to bring me to a state of happiness.

Exhausted

Burnt landscape

8

I Leave the Mountain

I walked out of the station and down to the road where Chris was waiting. I let Chris know what had just transpired. I was leaving Mid-Coast Fire Brigade. My days as a firefighter were over. So, we made a new plan. We decided to drive both trucks up to the property, drop one off, gather needed items together and head into town. We would stop by and check in on Rurik and Allie on the way up, and let them know we were heading out. If they needed a ride back to Highway 1, we would give it to them. Chris could head back home, and I would leave the canyon, knowing that my neighbors would watch my place.

I was still feverish and not feeling better. I needed to be somewhere cool and away from the stress of the fire. My friend Kierstyn had offered me a bed at her place in Carmel-by-the-Sea, which I felt was the refuge I needed.

We did not spend a long time at my place getting things ready, but had I known I would not return for almost two weeks, I would have stayed longer, put more things into order, and taken more with me. Chris gathered all his things while I loaded some bags of clothing, and a pillow.

We headed down the hill and picked up Rurik and Allie, who had been waiting and hoping we would come by to pick them up.

We drove to Highway 1 and parked behind Rurik's pick-up on the east side of the highway, near Grimes Ranch. It was incredibly soothing to breathe in the cool, moist, air, scented with aromas from the churning Pacific less than 100 yards away. Across the highway, on the ocean side, a News crew was preparing to film a woman reporting on the fire. She was attempting to put in order her superbly coiffed hair, while the persistent breeze negated all of her attempts to do so. She finally gave in to nature's nudging, and the crew got down to business filming. The scene was rather amusing to all of us, having just come off the burning mountain, with our soot-covered hands and faces, and ash-covered clothing. We wondered why they did not think to come over to our little crew and get the real story.

I spent a week managing all that comes with the threat of wildfire, both as a homeowner and a firefighter, but here we were, away from all that, in the cool ocean breeze. I pulled out a watermelon that needed to be eaten and we were soon sitting on the tailgate of Rurik's truck enjoying that cool sweet melon, the cool ocean breeze, the beautiful coast, and one another's company. Miraculously, our places were still standing and safe, and the fire seemed far away.

Watermelon finished, Chris and I made our relaxed way up the coast to Carmel. We hung out and talked for a little while at Kierstyn's, before Chris was on his way back to Fair Oaks. As he was leaving, he let me know that the last few days were the most intense he had ever lived through. I was glad to know those days had provided him with an out of the ordinary experience to add to his life story. I was sad to see him leave.

Amazingly, in the two days he left his truck in Carmel, he had not received a ticket for parking in a two-hour zone.

I soon made my way to a shower and into a bed at Kierstyn's. My fever was around 100 and manageable. My lungs and throat burned as did everything else inside. I felt like I had eaten too many hot peppers. That night, I awoke three times dreaming about fighting fire.

By the time the morning light made its way in through the windows, I felt as though I had not slept at all. I was restless, but lay still, hoping I might fall back asleep. Kierstyn was soon on her way to work, and I had the quiet house to myself. My fever continued, and remained around 100. In the early afternoon, I decided to try to do some things. I went to Martin's Irrigation in Sand City, to get water line for my neighbor Christian. The fires had burned his water system including a large solar array, hundreds of feet of waterline and a DC water pump.

Martin's three-generation business had been my source for plumbing supplies while installing water and irrigation systems. I was one of their regular customers and during the course of those many visits had become friends with the family. Frank and his son James wanted to hear about the fire, so I hung out for a while sharing my experiences with them.

I soon started to feel extremely fatigued, finished up our conversation, and headed a few blocks away to a self-service car wash to rid my truck of its patina of ash and road dust. The truck had become matte grey rather than its glossy black and red finish. Parking in one of the bays, I noticed it took quarters to operate the wash, which I did not have. I walked to a friend's auto parts store, half a block away for change. I started to have the same experience of a few days before when I went shopping at Safeway. Everything seemed to be in slow motion and doing anything at all seemed to take quite an effort. I felt like I was entering that same strange state I had been in when I drove back to the canyon and couldn't recognize where I was on Highway 1. I came back to the carwash with change and deposited the coins into the spray machine.

The moist mist created by the high-pressure sprayer was delightful to feel on my skin. Each breath of the moist air soothed my burning lungs. When I was finished, my truck looked clean. The interior was a different story. Everything inside was covered with fine ash, and the smell of smoke strongly lingered in the fabrics.

Though refreshed by the sprayer, I was steadily feeling worse. I started driving toward Carmel and Kierstyn's home, about six miles away via Highway 1. For some strange reason, I drove through Monterey, Pacific Grove, and Pebble Beach to get to Carmel. I thought the beauty of the drive and staying close to the cool ocean would do me good. The altered state of perception though, like being in a time warp, continued. It felt as though it was taking forever to get there, and that maybe I never would. When I finally got back to Carmel, my head ached and my fever was elevated—102.2°.

I lay down and tried to rest, but my head hurt so intensely I could not sleep. My thoughts were jumbled. I lay there in pain, feeling miserable. Kierstyn came home with her two children, Belle and Simon. Even their good humor and playful exuberance did not get me feeling any better.

The fever continued to increase and as it did, so did the headache. The pressure I felt in my head was unbearable. I wanted out of my body. It became clear, as the minutes ticked by, that something needed to be done to lower the temperature and reverse the symptoms that seemed to be getting worse. It was intense!

Kierstyn offered to take me to the emergency room. I suggested that we wait a little while to see if the fever started going down. We waited about 20 minutes and checked it again. It had risen to 102.9°. I asked for 20 more minutes.

"If it does not come down, I'll call 911. Maybe Carmel Fire Department will come over and give me some oxygen," I said hopefully.

Twenty minutes later my fever had risen to 103.5°. Not wanting to go to the hospital but wanting help, I called 911. The dispatch operator answered, "What is your emergency?"

I explained, "I'm a firefighter and have a high fever and bad headache. I just spent a week on the Soberanes fire. I'd just appreciate having Carmel Fire Department come and check on me, and make sure I'm OK. Could you please ask them to come down quietly—no lights or sirens? I am not dying or anything like that."

The operator responded, "I'll get them on the line."

Someone from Carmel Fire Department answered. I explained my situation to them, and they let me know they were on their way. Within a few minutes they arrived, quietly, as I had asked. They knocked on the door and Simon went down to let them in. Up the stairs they came, in their turn-out gear. There were four of them, and with their bulky uniforms they seemed to completely fill Kierstyn's small apartment space. They took my vital signs. They informed me that everything was normal, except my temperature, which was then 103.3°. Even my oxygen levels were good, which for some reason I was the most concerned about.

Belle and Simon looked on, amused. What a thing to have the Carmel Fire Department in their living room!

They would not give me oxygen. The fact is they would not give me anything except a ride to the hospital.

"Theo," a paramedic who knew me exclaimed, "I can guarantee, you will not sleep if you go to the hospital!"

After a moment of silence, she went on, "Theo, you have to quit being so 'Big Sur!' Herbs and tea are not going to do it this time. You were in a fire for a week. Your sinuses are going to give you a headache for days. Get some Nyquil cold remedy and get some rest."

Almost before the paramedic finished, Kierstyn was out the door and on her way to the local liquor store for Nyquil. Before the fire men and women left, Kierstyn was back with the goods. The helpful, friendly fire people from Carmel bid us all a goodnight.

I took two Nyquil and then thought about letting one absorb right through the soft tissue in my mouth for quicker results. I would have done almost anything to get away from the pain. I took a third one and bit into it, allowing the bitter fluid full access to my membranes.

Within minutes I was asleep and did not awaken for over four hours—the longest I had slept in days. When I did awaken it was still dark, but at least my headache was mostly gone and my body relaxed. What I found the nicest in the dark of that night, was experiencing the cool moist ocean air coming in through the open window.

During the next two days, I did not leave Kierstyn's house. I spent hours sleeping. My temperature fluctuated, but never again went over 102°.

By August 1, Monday morning, for the first time in ten days, my temperature was normal. I felt better, but delicate and weak. I decided to go check in at Mid-Coast Fire and to try to get water lines to Christian.

The drive down the coast was beautiful. It was a surprisingly clear day. The breeze had blown the smoke to the south. Turning up Palo Colorado Road, I was recognized by the Highway Patrolman blocking access to the canyon. He let me in and I drove up to Mid-Coast Fire Station. I spoke briefly with Cheryl, who would not allow me to drop off irrigation supplies to Christian. She wanted me to suit up to help out. I explained to her that I had just had a fever for ten days, but she seemed to think that was no excuse for not suiting up.

I got into my truck and left. I would not be coming back to the canyon until the public was allowed in.

It was difficult to drive away from the canyon, realizing it might be quite a while before I set foot on my property again. I had not prepared for being gone a long time. The garden irrigation system was not repaired and it was hot, dry weather. I had to accept that everything in my garden would be dead or close to it. I hardly had any clothing—one nice shirt, a couple pairs of pants, a few t-shirts and some socks and underwear.

Later that afternoon, I ended up at Happy Girl kitchen and mentioned to Todd Champagne, one of the owners, that the next day was my birthday. I expressed to him a wish to have a little celebration, and before I could say anything else, Todd offered Happy Girl as a place to gather. We created a Facebook Event and posted it. It wasn't much, just that I was having a birthday and to please come and help celebrate at Happy Girl from 6-8pm.

My Birthday

My birthday morning was spent sleeping as much as possible. I was still exhausted, both from the fire and the fever. Whenever I attempted to do anything, I soon just felt like stopping and lying down. There was nothing pressing to do, and there was no getting to my place, so sleeping was a healing luxury that could actually be realized.

Arriving at Happy Girl a few minutes past six there were already people there. Two friends were helping Todd prepare some food.

It was an amazing evening with the most incredible friends showing up.

Many wanted to hear about the fire, so I shared some stories. One friend and longtime Big Sur resident expressed how angry he was at whoever it was that had made the illegal campfire resulting in the fire. He asked how I was dealing with that.

I let him know that I too had been very angry at first and had been all riled up to do all I could to bring awareness to the problems tourism presented to the Big Sur community. Then, I shared the dream of Agni, and my acceptance that this fire was inevitable—especially since our forests are not managed in ways that minimize fuel loads. How the fire started did not matter—it was going to start one way or another. A fire would have eventually burned through the Palo Colorado Canyon community and the surrounding area.

"We all had time to get ready and prepare our properties for years. Fire is a part of the California landscape. Without fire being a regular feature of the landscape, fuel loads build up. When there is a fire, it is disastrous. We need to mimic the effect of fire on our properties to keep them vibrant and safe. We would be wise to adopt the practices of the indigenous Californians and burn regularly."

He appreciated all I shared and visibly relaxed.

The party was over around 9pm, and after beginning to help clean up, I was shooed away by Todd. I started on my way back to Kierstyn's, feeling thankful. As I did, I felt an urge to try another approach to getting back onto my property to take care of things weighing on my mind.

Every birthday I like doing something adventurous or unusual that will help me remember the day. Not having planned anything except the party for my birthday, I thought, *Why not hike up to my property from Highway 1?* I figured this fell into the adventurous/unusual category.

I could park at a pullout on the highway, hike down an access road, get onto a trail to cross Rocky Creek, then up driveways, a trail, and a fire road to the Hoist on Palo Colorado Road. From the hoist, I could hike up Long Ridge Road to my house.

I had the irrigation supplies I needed to repair my irrigation system which could hopefully save the more delicate plants in the gardens and orchards. I could also get some more clothing.

I arrived at Kierstyn's and we talked for a while. She had come to the party and been amazed, like myself, at the wonderful group of people. I told her of my intention to get up Long Ridge. She thought it funny, but adventurous, and never tried to persuade me otherwise.

After midnight I started off south along the coast and parked at a pullout near Rocky Creek Bridge on Highway 1. Filling a backpack with irrigation supplies, and locking my car, I disappeared into the dark, down a dirt road toward Rocky Creek. Though I had a great headlight, I did not want to be seen, so I kept it off. What I was doing was not sanctioned for homeowners—entering an area where a mandatory evacuation was in force. I doubted anyone would care, but I thought it best to stay as incognito as possible.

Though I knew there was an obscure trail across Rocky Creek, it took me half an hour to find it. Once on the trail, I was able to cross the creek to the other side. On the other side was a private paved road that came down to Rocky Creek from the Old Coast Road. I turned on my headlamp for a moment and was startled to find a large fresh Mountain Lion scat in the middle of the road.

Briefly, I reflected on how amazing it was to be trekking along in a diminished zone of green, surrounded by burnt landscape, through mountain lion territory, in almost complete darkness. It was a new moon and foggy. Not even the light of the stars was there to illuminate the terrain.

Because of the fog layer, the air was cool and damp; a great temperature for the exertion that comes with a steady elevation gain.

Occasionally, I needed to turn on my light to see where I was going, but mostly I kept it off. I had traversed over this land enough to be quite familiar with it.

I hiked up the paved road toward the Old Coast Road, and could barely make out two fire trucks ahead, parked on the Old Coast Road for the night. To avoid walking by the fire trucks, I went up a small game trail using my hands to feel the compacted earth. The game trail led me onto a dirt driveway/fire road that ascended Long Ridge from the Old Coast Road. Once around the first bend, and out of the line of sight of the fire trucks, I turned on my light.

The road was soft, like Long Ridge, turned into a light fluffy dust from bulldozers and fire engines. It was slow going, because with every step my foot slid back a little. I tried to stay toward the edge of the road where the ground was more solid. There were fire hoses running along the road, empty water bottles, and wrappers from firemen's snacks.

I kept going up and began traveling through the layer of fog. It became more cool and extremely damp. Water drops formed on my bare skin.

It was slow going and strenuous, especially after the extended fever.

After plodding along for quite a while, I eventually came to where the ridge leveled out for a little way. I had hiked above the fog and was amazed to see the glow from the fires far to the south and east, and stars overhead. In the distance I could see a fire in the road. That seemed odd. I got closer and realized there was a person standing next to it. A group of firefighters were sleeping around the fire. They had started the fire to keep warm through the cold night.

Not wanting to interact with anyone, I took a trail that passed above where the firefighters had made their camp. I did not want to have someone tell me to go back.

Climbing past the level area and continuing the ascent, I reached an area where the ridge had been transformed. What was once a wide expanse of brush with a trail running through it, had now become an 80-foot-wide dozer scrape. I was shocked and stood there taking in the altered landscape.

The trail that ran through this area was gone. Much of the trail had been overgrown and almost impassable, but it was nice that way. It was used mostly by wildlife and the occasional local. I liked mountain biking down it and had done so regularly.

I started up a very steep section of this newly bulldozed expanse. Having gone no more than a mile I realized that above me, on a flattened area, was a fire engine. I would have to walk right by it. I wondered at first how it possibly could have gotten there, but realized the bulldozers must have cleared a road for it from the Hoist.

I did not like the idea of walking past that engine and stood wondering what to do next. It seemed the way forward was blocked. There was at this point one way up, and it would be right past the fire engine. I decided not to go any further—not wanting to make trouble for myself or anyone else.

I took in the whole scene. From that vantage point, the glow of fire off in the distance could be seen in many parts of the landscape. It was beautiful to witness—the orangish lively glow. A sweet calm came over me. There was nothing more to do. I had tried, to the best of my ability to get home, and wouldn't make it. I surrendered. I would get home when the community was allowed back into the canyon, and until then I had a sense that all would be fine.

With a light heart, I turned around and started down the hill.

One thing for certain was the joy I felt with this wonderful nighttime adventure—one I would remember for a long time.

A couple hours later I arrived back at my car. Like so many other nights recently, the sky was getting light in the east and I was still awake. I looked back the way I had come and saw that the fire crew, who had slept on the Old Coast Road, were awake and stirring. The lights on their trucks were on and people were milling about.

The drive up the coast in the morning light was beautiful. Arriving back in Carmel, Kierstyn was just getting up. I made her hot water and then, her morning cup of coffee. I had some coffee, too, in hopes that I could start my day as if I had not spent the entire night up. I sat down with Kierstyn, but the moment I leaned back on some pillows, I was asleep.

The next days were a blur. I slept a lot. Though I had plans to accomplish some needed tasks in town, it seemed I did very little. The present moment was all that held my focus and being so exhausted, sleep was the easiest and most natural thing to do.

Theo surveying gardens and orchards after the fire

9

Returning Home

Tuesday, August 9, 2016, the community was allowed back into the canyon. I went home, but I was not excited to return. It was smoky, the landscape was burned and altered, and I felt a deep sadness thinking about the loss and destruction. 57 homes burned, my cherished madrone forest burned, and who knew how many wild creatures lost their lives. Everything looked and felt different. The community would never be the same. Yes, I had those dreams. I could embrace that the old needed to go to make way for the new, and that change is what keeps everything beautiful, but to see the beautiful madrone forest that graced the north side of the property adorned now with brown dead leaves, I was brought to tears. Thinking of the wildlife that could not flee in time brought on a sense of hopelessness. Many of my friends and neighbors had lost so much, and in some cases everything. Many people were trying to figure out where to begin. I shared that dilemma.

That day, I drove down my lower roads to see what the fire had done. My storage area had burned. 17 telephone poles were gone. A pile of steel roofing had all but disappeared. Everything that could burn had become ash, and a surprisingly thin layer of it.

Everything in the metal storage tank was incinerated. Before I bought the property, the 3000-gallon steel water tank had been used as target practice, leaving bullet holes through the steel sides. Knowing the tank could not be used for water, I had transported it to a storage area on the property and cut a steel door in the side.

Four tires on aluminum rims had been stored within that tank. I had listened to their explosions with the CalFire chief when the fire came up the hill from the north. The explosions had distorted the side of the water tank—perhaps irreparably. Nothing much remained of the tires and rims inside the tank, except the steel bands from the tires

Theo, pointing toward his property, and island of green

Theo surveying gardens and orchards after the fire

Inspecting two burned cars

Sifting through the burnt wreckage

A yucca plant blooms in the burnt landscape (see opposite page)

What is left of a 2500 gallon poly water tank

New life
(Photo by Michelle Magdalena)

and three of four rims melted into very small puddles of aluminum. Aluminum melts at 1241° Fahrenheit. That's hot. Everything plastic inside the water tank was gone.

As I looked into the tank, I saw some long, four-inch-wide bands of fiberglass running parallel to some thin steel strips. I wondered what these were and after some time contemplating the remnants, realized they had once been downhill skis.

Four redwood burls were gone, as well as a nice stack of old growth redwood lumber I had saved for some as yet unknown future project. A stack of sliding glass doors for a greenhouse had broken and melted.

A used Solahart solar hot water heater had burnt beyond recognition.

A 30' boat trailer was still intact but without lights, electrical wires, or tires. Where the tires had burned the frame had distorted. Two old Mercedes Benz 300TD parts cars were gloriously burned. Mercedes unique seat coverings had provided ample fuel to get the inside of both cars sufficiently hot to burn everything away. Every pane of glass had melted making unusual sculptures on the steering wheels, seats and floors. The frame of one of them had collapsed where it was supported on concrete blocks.

A 2500-gallon poly water tank had become a flat pancake of green plastic.

Fourteen cords of stacked and split firewood had been reduced to a four-inch layer of ash. I had held on to some irrational hope that this large pile of winter income somehow miraculously escaped being burned.

Other than the thousands of feet of pvc waterline I had run through the woods, there was little else of material value that had burned.

Worse for me than losing all those things, was not knowing the extent of the damage to the trees. The madrone forest on my land was more beautiful than any other madrone forest I had ever seen. It had looked like a well-maintained park. The trees were 50-60 years old and we spent a great amount of time clearing the dead trees from the hillsides, in an attempt to make the landscape more fire safe. The last fire had been in 1964.

I made my way back to the house. Feelings overwhelmed me. I was sad seeing the altered landscape, thinking it a wasteful destruction. I was angry that many plants I had nurtured from cuttings were burned.

I came inside my house and sat down at the kitchen table, unable to figure out where to begin with restoration, clean-up, and trying to live into a new sense of norm.

Everything seemed safe and secure around my buildings—an island of green in the middle of a transformed landscape. As I walked over the land the views beyond the green zone from my house and property were of a grey landscape, with the vegetation in many areas along the south sloping mountain completely burned away. Most of what had once been manzanita bushes had become fire-blackened branches devoid of leaves. The forests on the north sides of the mountains were a canopy of brown leaves, and though there were islands of unburned trees, in most places the layer of leaves and organic matter making up the forest floor had burned away completely, replaced by a fine layer of ash.

The smell of smoke pervaded everything.

The whole experience of the fire had been overwhelming and intense—made more so having had a fever through all the days the fire made its way through the terrain around the property. There had been so much transformation and so much destruction, and it happened so quickly. I was awed by the power of fire and its ability to transform not only the landscape but also my state of being.

I had a better understanding for what it would be like to survive a war. There is nothing one can do to bring things back to the way they were. It would have been best had I been able to simply accept the new state things were in, but I found myself resisting it, and wishing things had not happened the way they did. I was grateful for having made it through the fire with all of my buildings intact, and the gardens and orchards only slightly altered. It was in fact, miraculous. The way things had unfolded was miraculous, but could I keep my attention on the positive?

Being at my house pushed me into a negative space. It was too much to adjust to. I found myself wanting to get away from the whole scene. The property was safe. The fire was far away. The horses were at the SPCA. The gardens were again mostly on a drip system. There was no need to be there. I came to the conclusion that getting away could be a good thing. Some unburned landscape and some clear air to breathe might help me readjust my perception.

Life continues

10

Surrender

After leaving the property and the canyon I did not spend a night at home for 10 days. Time was essential to process all I had been through, as well as being somewhere other than what looked like a war zone. Mostly I stayed in Carmel at Kierstyn's. During that time, a group of us put on a creative event in Sand City for Community Palette, a non-profit that raised money for local artists and businesses. That year the funds we raised went to people who had suffered loss from the Soberanes Fire. With that as a focus, my mind slowly came out of what is known as acute stress disorder.[2]

While working on set design and construction, I was able to interact with people who had had nothing to do with the fire. In other words, conversation was not about the fire, giving me a chance to think about other things.

I had simply been too fully 'on' for too long. Responding to fires and emergency situations as a firefighter usually put me into a state of hyperarousal, but it also was usually over in a matter of hours or, for fires, a couple days. In many ways, being in a state of hyperarousal is a wonderful state to be in. Everything I would normally be thinking drops away. I end up being fully present in the moment—in the now. Functioning from that state for extended periods of time though, taxes many systems in the body to the point of overall exhaustion. This can then lead to PTSD—Post Traumatic Stress Syndrome.

Days later, on my return home, I did my best to positively take in the new look of the property, the landscape, and the community. My estimate was that the fire had burned through 37-38 acres of the property, leaving a 6-7-acre island of unburned green.

---

[2] Acute stress disorder (ASD) is a short-term mental health condition that can occur within the first month after experiencing a traumatic event. It involves stress responses, including, anxiety, intense fear or helplessness, flashbacks, or nightmares.

On Long Ridge, there were another ten homes, and the areas around them that had not burned. All of this became a small oasis of life, sustaining the small mammals and birds that had remained on the land through the whole ordeal. Their sweet songs and lively chatter filled the days and nights with the evidence of life, though there were not as many lending their voices to the symphony.

Some days, awakening with a lively wonder, I anticipated what I could do or accomplish in the day. Just as often my thoughts would move toward the destruction I perceived had happened, usually sending me into a loop of negative thinking.

Though things at the property seemed good, and once I moved into activity, I tended to feel OK, I was still having nightmares about the fire, almost every night. I dreamt about backing my water trailer into the fire, standing on the edge of the forest on fire, arguing with people about the fire, working the fire with a shovel, all of them accompanied by chaotic jumbles of images leaving me feeling like I was still in a heightened danger zone—still dealing with the fire. It always took time to realize it was a dream. I had never experienced anything like it. Falling back to sleep the dream would just start right up again. The only way to stop the dream from continuing was to get up and fully awaken—not a recipe for good rest.

The dreams kept coming and some became repetitious. It seemed I was stuck processing through the experience of the fire. I started waking up in the mornings wondering what was the point of getting up or doing anything—everything could change for the worse in a moment. If I laid in bed, it only intensified.

This was not something completely expected. It is true that it took a while for me to process every major fire or medical incident I responded to as a firefighter, but this was different. Before it was someone else experiencing the nightmare. All those other times I could distance myself from it, because I did not have to deal with it after the event was over. This time, the nightmare was mine, and I was living with the results of the fire every day.

The overwhelming feelings of loss and despair were difficult to experience and feel. The world was so transitory. Everything in it could radically change in a second. Everything I had worked to create on the land could have been consumed by that fire.

Could I transcend it all? Wasn't this the perfect situation from which to awaken into a state of non-attached enlightenment? Wasn't this the kind of event that allowed people to instantly gain a light-filled state? Wasn't this simply the old falling away to make way for the new? Wasn't this the change that kept it all beautiful? Maybe it was, but I just couldn't seem to get there.

The thoughts that kept coming were that life was very much like creating a sand mandala. You work very hard creating something beautiful, yet at any moment the winds of change can come and blow your creation away, transmuting its beauty into an undefined layer of dust. As with a sand mandala, it is important to relinquish any attachment for permanence. All we have done cannot and will not last for long. As is always the case, the time comes when everything changes.

Though my rational mind could entertain these thoughts, as evidenced by dreams, the subconscious had another story to tell.

Jennifer Allen, a friend in the Palo Colorado Canyon community who specialized in PTSD, offered to do a few sessions with me. She had worked with me after the Pfeiffer fire, getting me through the PTSD I experienced then.

In one of my first meetings with Jennifer, it became clear that the PTSD was a culmination of many unresolved traumatic events experienced during life. When too much trauma is stored in the mind and body, the result is inevitable.

Little by little over the weeks I met with Jennifer, my focus shifted toward more positive thinking and action. Still, struggling with depression made it difficult to orient to something meaningful. I had to accept that my place, which I found so heavenly, was forever changing, and that as much as I wanted to hold onto it in a state I liked, that was not possible. It was reckless to base happiness and sorrow on things and situations that were forever shifting. Was there a place I could experience true peace and joy?

Remembering my practice of Transcendental Meditation, I knew I had perhaps the most effective tool to shift and heal—transcending. Meditation became a twice daily practice, and at least during those times there was relief from the more negative thoughts and feelings. The words of Maharishi floated into awareness.

> "We must take situations as they are. We must only change our mental attitudes toward them."

> "Rest and activity are the steps to progress."

And,

> "Do not fight the darkness. Let the light in and the darkness will disappear."

The best way to change my mental attitude was to spend time in the deep silence that lies within and doing TM allowed access to that. That brings the light to the darkness.

Then coming back into activity, I had a more positive orientation.

Focusing on two things helped. First, doing things I took pleasure in and was naturally good at that brought me joy to simply do, without being attached to the outcome. Second, as much as possible, to stay in a state of love and acceptance—to be loving to myself and others, and to lovingly surrender to and accept change. More and more that became a choice.

Things had unfolded in a fortuitous way. Even the 102° fever had been a fortunate condition allowing me the freedom to stay at home, rather than respond to the fire as a firefighter. Had I responded as a firefighter, I would likely have been ordered to be present somewhere besides my property, and the results for it, and the rest of Long Ridge probably would have been disastrous.

Many people lost everything. I made it through the fire with all my structures, gardens, and orchards intact and still had an island of 'green' unburned landscape. There was a lot to appreciate.

The goal became to put energy and creativity into what there was to work with—buildings, gardens and orchards. Also, to select the thoughts I wished to enliven. Working to bring beauty to the outer surroundings, as much as possible, I choose to keep thinking positively. Managing a large piece of land, the work is never ending. Things are always growing and changing through the seasons. Spending as much time as I could working outside as a good land steward, I felt held by nature's rhythms and nature's laws.

11

Living with the Aftermath

Everything had changed for our coastal community. A huge expanse of landscape was burned—132,127 acres. In the Palo Colorado community, 57 homes had burned. Some people had evacuated so quickly that they lost everything they once called their own. Others had been away for summer excursions and unable to do anything but watch from afar. Residents were kept out of the canyon for over two weeks and most, when they did return, took time to orient to the changes wrought by the fire. Most tenants whose places had burned never returned to live in the community. They might return to sift through the burned remains of their previous dwelling, but for most the fire created a *no return* scenario. What had been such a sweet, positive, joyous community of people was forever altered.

On Friday, July 22, when the fire started and it was clearly making its way to the Palo Colorado community, a mandatory evacuation was issued. This was communicated to everyone through a community phone tree. Most people took it seriously. Anyone who lived in Big Sur long enough knew that summer wildfires were intense and unpredictable. Wisdom and experience would inform any sensible person to get out and head somewhere safe.

Laura Jensen, the head of our phone tree, had called me and let me know they were leaving as soon as they were able. Laura and her husband Dru had lived on Long Ridge for most of their lives. Both their children were born on the ridge, and they were not strangers to the hazards of Big Sur wildfires. They knew the safest option was to leave.

The Palo Colorado community has one road for getting in and out—Palo Colorado Road. It is narrow, and in many places only one car or truck can pass at a time. There are numerous sharp, blind curves making it difficult to maneuver sizeable vehicles up and down the road.

In the 19 years I have lived in the community, there have been four separate fire events during which ingress was restricted to fire personnel and law enforcement. To bring emergency response equipment up that narrow curvy road, there cannot be interference from residents trying to drive in and out, especially in an elevated emotional state.

In response to a fire like the Soberanes Fire, a wide variety of resources are requested by the Incident Commander. During the week when the properties of the Palo Colorado community were threatened, there was a steady stream of emergency vehicles (fire engines of all shapes and sizes, water tenders, bulldozers, bulldozer transports, ambulances, fire chief vehicles, fuel trucks, equipment vehicles) going up and down the main road, as well as up and down the many side roads that led deeper into the rugged coastal landscape where most people had created their homes. Almost all the side roads were dirt/gravel roads necessitating 4x4 fire engines.

By the time the fire was deemed non-threatening, and firefighting equipment moved out of the community and closer to the moving fire front, the once well compacted dirt roads had turned into many inches of fluffy dust-like material. Going up and down the roads was slowed by decreased traction. By the time you arrived either at the top or bottom of those dirt roads, your vehicle was covered by fine dust.

The paved Palo Colorado Road had been traversed repeatedly by bulldozers. The road was scarred by their steel tracks. The sound these scars made as tires rolled over them reminded me each time of what we all had just been through.

Though the lower part of the canyon had not burned, once above an elevation of 800 feet much of the landscape had. It now shown in varied tones of green and tan, to shades of grey and brown. On south facing hillsides most of the organic material had burned. Twisty, blackened manzanita branches looked like skeletons reaching up out of the dreary, ash-covered ground. It looked like a graveyard and in fact in many ways was.

Northern slopes once hosting the many species of California coastal trees stood now as stark witnesses to the fire's intensity. Where the fire was the most intense, some trees had completely burned. Redwood trees, adapted to fire events, remained standing but often with their branches and bark singed completely. Madrone trees with their thin bark, were burned through their lower cambium layer, rendering the top portions of the trees dead. Those trees were now hung with dead leaves transitioning from green to tan and brown.

Though fire can have a beneficial influence on the landscape, when fires burn too hot there are more destructive impacts. Much of the landscape the Soberanes fire ripped through had not burned for over fifty years and fuel loads had built up, giving the fire plenty from which to burn hot. In many places 50-60-year-old forests were burned severely, to the point where recovery would take a long time, and it would, of course, never be the same.

There was nothing to do besides accept the new. As much as I could, I looked at it from a neutral place—neither good nor bad. It just was. I did have a lot to be grateful for and found myself dropping many of my attachments to 'my' place. There was freedom in that.

What to do next occupied my mind. There was so much to do, it was just a matter of prioritizing.

Much had been lost and burned in a lower storage area. There was also destruction near my home. During backfire operations, CalFire crews burned 400 feet of netting that encircled gardens, a portion of the orchard, and much of the irrigation system. A bulldozer had made a couple of scrapes close to the house and barn. Hundreds of feet of installed waterline had burned down in the woods.

Before working on all that, I decided to check in with the insurance company to see if some relief was available. They connected me with an adjuster who soon came for a visit. We surveyed the fire's aftermath. He created a list of all the things that had burned, and surveyed my house and structures.

"I'm going to take care of you, Theo," was his response to me, and he did. He valued the lost possessions and damage to my structures due to smoke at a little over 50K, and a check was soon in the mail. This helped, a lot. I suddenly felt like I could relax a little. I could focus on needs around my property and clean things up, as opposed to worrying about all that and working to make money.

Soon after, the insurance company dropped me. They also dropped another neighbor they had insured just up the road from me, whose place most likely would have burned had I not stayed and stopped the fire. I found it a little frustrating. In reality, I saved that insurance company hundreds of thousands of dollars by staying and defending the ridge. I kept all of my structures from burning, and probably my neighbor's. But that would never be considered. Like so many other people all over the USA living in fire prone areas, house insurance has become more and more difficult to acquire.

When the insurance company let me know I was being dropped, I ended up creating a policy with California Fair Plan for less coverage, but at a rate 50% more than what I paid before. I thought about dropping insurance all together, depending on my ability to defend my own property from fire, but life is fickle. There is no guarantee I will be home the next time there is a major fire.

Cleaning up my property entailed loading burned materials onto one of my trucks and making a few trips to the dump. It was extremely dirty dusty work, best done with a respirator and protective clothing. The ash was extremely fine, and much that burned consisted of manmade materials. A Mennonite group offered to come to people's properties and clean up the aftermath of the fire, but I opted to decline, thinking there were people that needed that limited service more than I did.

I lived into the burnt landscape. It was fascinating, with its varied hues of grey and brown. If there was ever a time when a person could actually see the contours of the landscape this was the time. In most places the ground had been burned completely of all organic matter and vegetation, resulting in the 'ground' level dropping by inches. There were still trees, but for so many the top portion had been killed—the fire having burned through the cambium layer.

Having nothing on the ground made travel through the forest easy, but with this came the concern that when the winter rains commenced, the steep mountainous terrain of Coastal Big Sur would facilitate water and debris runoff at an accelerated rate, potentially having disastrous consequences for everything in proximity to a riparian zone.

12

Devastating Winter Storms

It was a winter with normal rainfall, but not normal landscape or road conditions. As predicted, when the winter storms delivering the most water arrived, the streams became choked with sediment and debris. As the rain fell at its usual fast, coastal rate, the top layer of burned soil and ash quickly mixed with water and became a powerful abrasive, exponentially adding to the amount of sediment carried into the streams, and creating a force able to move objects of great size and weight. Branches, logs, boulders, and even whole trees were carried in the flow. Culverts along the length of Palo Colorado Road clogged. Once the culverts were blocked, the water found the path of least resistance, which was typically over the part of the road that the culvert was under.

Long Ridge has watersheds on either side. On the south side, Turner Creek runs into Mill Creek, which makes its way to Bixby Creek and out to the Pacific Ocean. On the north side Rocky Creek runs the length of the ridge, making its way to the Pacific Ocean a mere quarter mile northwest of Bixby Creek. As Palo Colorado Road wends its way inland from the coast, it runs along Palo Colorado Creek and passes over it many times. It then ascends over a ridge at Murray Grade, and makes its way down toward Rocky Creek, crossing it before continuing up to the Hoist, where Long Ridge Road begins.

The four-foot culvert in place under Palo Colorado Road allowing Rocky Creek to flow beneath the road became severely blocked. Despite the best efforts of residents with backhoes and tractors it could not be cleared. Some water was getting through. but not enough to keep the water from collecting on the upstream side. The water was already too deep to get to the blocked culvert. The water continued collecting upstream from the culvert until it began to cascade across the road. The volume of water moving down stream was so great that it began eroding the ground away on the upstream side of the road. At a certain point it began to flow between layers of asphalt. Once begun, huge sections of road began to peel away. Rocky Creek was unrelenting. Within minutes the creek had moved a significant section of asphalt completely away, leaving the base rock beneath vulnerable to its abrasive onslaught. Eroding the base rock away took very little time at all, after which the water flow carved a deep passageway through the softer material. A ten-foot section of road had disappeared as Rocky Creek carved its new watercourse. Palo Colorado Road had become impassable.

It was impressive to see the power water has to quickly alter the landscape, but those of us living beyond the washout had a new situation to contend with, and no easy way out either in a vehicle or on foot.

Most of us were prepared for an event such as this. Living in Big Sur was unpredictable. The steep terrain, so close to the Pacific Ocean, created a dynamic, quickly changing biosphere. Temperature fluctuations in a given day could be extreme. The landscape could go from being encased in low lying clouds or fog to being in the brilliant sun in seconds. It could be raining or snowing on the mountain tops, while sunny and warm at the coast. Storms blowing in off the Pacific could deliver massive amounts of rain in hours. Wind velocities on ridges often exceeded 100 MPH.

Most of us had gardens and orchards and food supplies for both people and pets for a couple months. The response from most was a celebratory one of having a nature-imposed, time-off from the normal schedule.

The county's response to this minor disaster was quicker than anyone expected. Our fire chief, Cheryl Goetz, represented the community, demanding that a temporary bridge be put in place for emergency purposes.

Trucks deposited loads of large rocks and boulders, as well as 20-foot lengths of 8-foot wide laminated wooden roads. Large equipment was soon poised at the site and a temporary bridge put in place over Rocky Creek's new stream bed.

For the next few months as the winter storms waxed and waned, the stream's flow was carefully monitored. When an especially big storm was predicted, the wooden tracks of the bridge would be removed with a crane until the threat of them being washed away was over. These periodic road closures prompted many of us to leave a vehicle on the other side of the bridge, but that still necessitated crossing Rocky Creek. For this, a small wooden foot bridge was put in place over the creek in a stand of redwoods. Miraculously, it stayed in place through the winter.

Concurrently, other parts of Palo Colorado Road were suffering from blocked culverts, requiring removal of the material blocking them. Most of this work was done by residents—often done by hand, but just as often with backhoe and tractor assist.

That February, I had a trip planned requiring airline travel. The afternoon before my departure date, I was driving up the canyon (Palo Colorado Road) and got to a place where the road was blockaded. Just ahead of where I was forced to stop, water was running over the road because of a blocked culvert.

I backed up and parked my truck in safety, at a friend's place. I walked up the road to offer help. It seemed all was handled, but when I arrived on scene, Jake Goetz, who was on the other side of the blockade, offered to give me a ride to my other vehicle, about a mile away. The timing was perfect. I hopped in and made it home almost as fast as if I had driven there directly.

The next day, in order to get out of the canyon, and off to the airport, I loaded a bike into one vehicle at home, parked that vehicle in a safe place along the road, and then biked across the blockaded and flooded parts of the road. Much of the way the water was a foot deep, but sometimes even deeper. While crossing those deep sections I would take both feet off the pedals, balancing on the seat and hoping I didn't hit a pothole. It was a little nerve racking as I had strapped my luggage on the bike rack and was wearing my carry-on on my back. I could not see down through the murky water. During that adventurous ride, I hoped that even if there was something unexpected, I would manage to stay upright.

Somehow, making it down the road and across all the flooded areas, I arrived at the truck I left at my neighbor's the night before. Even with the unusual commute I had enough time to make it to the airport, and plenty of time to catch the flight.

Had I driven up Palo Colorado Road five minutes sooner the previous day, there would have been no problem getting through, but I never could have made it out as the road remained closed. There was too much water flowing over the road where the road had been blocked off. What originally appeared as an inconvenience, turned out to be an incredible blessing.

Returning one week later, I was able to drive all the way home as the rains had temporarily subsided. By then the county had come up with a plan to rebuild the blocked culvert. They would begin in July 2017, with the intent to be finished with the work by October and before the winter rains arrived. During that time, we would not be able to drive cars or trucks around or through the construction site. There would however, be a footbridge put in place to walk across, but also wide enough for a quad.

New sprouts coming up from the root systems of the madrones in the spring of the morel mushroom bloom

13

Nature Delivers a Special Treat

The winter rains were over, and though there would still be spring showers in April, the big storms were past.

The fire and the rains had cleansed the land and also the community. The fire burned things up, and then the rains washed the dust and ash away. On most properties, in the aftermath, cleaning up was necessary. For many people, their properties had been completely burned and what was left was charred remains. The county placed dumpsters along Palo Colorado Road in a couple of places to be filled by residents with any and all burnt materials.

Truckload after truckload of burnt debris came out of the mountains. Trucks would dump the burnt material in the road, where it was then rudimentarily sorted and placed into the appropriate dumpster container with a loader. The loader had been left by the county and there were a few people in the community who were designated and trained to run it.

Even though the area around my structures had not burned, I felt motivated to get rid of things not needed or wanted. It was like joining the mood of the fire—make way for the new by getting rid of the old.

While cleaning up one day, my neighbor, Pat Pallastrini, was walking on the road below. I called out, "Hey, Pat."

He looked up, and responded with excitement in his voice, "Theo, you cannot believe what I've found. I have been waiting for this moment forever. Come down!"

Consenting, I made my way down a trail to the road.

"Look at this," Pat exclaimed as he pulled a paper grocery bag out of his backpack.

I had no idea what was about to be revealed, but I knew it would be something interesting. Pat had grown up on Long Ridge, on the property next to mine, and was very much one with the land.

He opened the bag and I looked in. It was filled with dark brown morel mushrooms!

"I knew it," Pat said. "I knew after the fire there was going to be a bloom of morels, but not like this! They're everywhere!"

I asked Pat where he found them, and he replied that they were all over my property in the burned madrone forest.

"Theo," Pat went on, "these mushrooms go for $20-$30 per pound. We could pick hundreds of pounds and sell them to all the restaurants. They love morel mushrooms!"

I was in, immediately.

"Let's do it!" I responded.

We agreed to meet back on the road in 15 minutes—enough time for me to get grocery bags, a backpack, and some water.

We reconvened and dropped off the road down the hillside into the burnt madrone forest on my property. We immediately began seeing morel mushrooms popping up out of the bare burned soil. They were often in clusters close to the trunks of the madrones.

It was like being a kid at an easter egg hunt, and it did not take long to fill a grocery bag with mushrooms. We left the filled bags behind to be collected on the way back up the hill on our return to the road. It was easy to walk through the landscape because everything was burned away. That also made it easy to see the mushrooms from a distance, since their color was so obvious against the ashen grey of the soil. Pat and I worked the landscape together, usually keeping in sight of each other, but with enough distance so we did not overlap the area we were covering.

What a beautiful way to spend time. We talked a little as we went along, and often commented on our incredible finds. It was magic.

Within a couple hours we had filled all our bags and decided to return with our haul. As we hiked up the hill, we began to estimate how much money we could make selling what we could so quickly harvest. We were excited not only to make some dollars, but also to have a reason to traipse over the beautiful landscape we called home, while foraging for mushrooms.

We needed to figure out where to put the mushrooms for safe keeping. I did not have anyone renting from me at that point and the communal kitchen was empty. We brought in some stainless racks and placed the

Theo and Pat's beautiful morel mushroom harvest

mushrooms on those. In this way they had air all around and would not be likely to go bad. We estimated to have picked about 40 pounds of mushrooms that first day.

Over the next few days, we expanded our search to an area comprising at least 120 acres. We went all over the place hunting them in the burned forest and got pretty good at knowing where the patches of mushrooms would be. Typically, they were around the bases of big madrone trees. I had never experienced anything quite like it—foraging through the forest for hours. Pat and I would enter into a state of such presence. Often, we did not speak to each other for long periods, but flowed together through the landscape.

At the end of the fourth day of picking we had close to 200 pounds of mushrooms. The realization set in—we needed to start selling our harvest or the mushrooms would go bad. Pat was focused on hunting for the mushrooms, and though this was clearly the most enjoyable, I began the process of calling restaurants and setting up meetings with the chefs and owners.

I happened to have perfect boxes for packing the mushrooms. They were flattish and held 5 pounds of mushrooms.

The next day I made my first trip into town with boxes filled with mushrooms. I mostly went to restaurants where I had made an appointment, but also stopped at restaurants in the Carmel and Monterey area, even without calling ahead of time. The mushrooms sold themselves. Every restaurant took one or two boxes and within a few days were asking for more. All of them started having specials with local morel mushrooms.

We were selling the 5-pound boxes for $100 each, and soon had a nice customer base. Unfortunately, our business spree only went on for two weeks. By then the morel bloom had passed—there were no more to pick. By the end of our spree, we sold over 200 pounds of morels at a premium price, and another 100 pounds that had gotten a little old for less. In those two weeks we made a little over $5000, which we split. The bonus was the time spent in the woods.

There was however one additional highlight. While our morel mushroom business was booming, the EcoFarm conference at Asilomar Conference Center was also happening. Passing by Asilomar after dropping off mushrooms at a nearby restaurant, I stopped to check out the vendor tent. There were always books, seeds, and other farm related items for sale there. While talking with a friend in one of the common areas, I met Paul Stamets. Of course, we talked about the post fire morel mushroom bloom. He became very interested. He asked that I send him some samples. He was quite certain it was a unique species, found only in madrone forests. I did send him a small box of mushrooms, but never did follow up to see whether he found our mushrooms to be a unique genetic strain of morel. Maybe he just cooked them up in a batch of scrambled eggs.

The wonderful world of redwoods to walk through
to get from one car to another

A hillside that has had the brush cleared and dead trees removed

14

Culvert Construction

The spring and first weeks of summer quickly passed. As the start date for construction of the new culvert loomed ahead, all of us on the far side of the construction area bought needed food and supplies. A day or two before the start date, I drove a couple vehicles to the other side of the construction zone, so I'd still be able to drive in and out of town.

The temporary bridge was the first thing to be removed, which meant there was no longer an open road to drive from home, down the canyon and beyond. There was now a quarter to a half mile to walk to get from one vehicle to another, depending on where you were able to park cars. From that day until construction was finished, I grew to enjoy the walk from one side of the construction site to the other, and from one vehicle to another. Even though most of the way was on pavement, it was like a walk in a beautiful park, much of the way beneath towering redwoods. Adding to the beauty of the experience, when crossing the foot bridge, there was the gentle melody of Rocky Creek as it cascaded over rocks and boulders on its way to the Pacific.

A few weeks before construction began, I bought a used Honda quad specifically for hauling groceries or needed supplies from one side to the other. I built a wooden platform on the back rack and secured plastic, rectangular baskets to the front racks. This enabled me to carry a great number of things safely from one side to the other. Though I had never imagined owning or using a quad, it turned out to be extremely helpful and was fun to drive. Its only flaw was two tires with slow leaks, requiring pumping every time I drove it.

I had taken on the responsibility of another dog. Chris (my son) and Linsey (girlfriend at the time, now his wife), had huskies with a litter of pups. I ended up with their favorite pup.

The dogs accompanied me on most of my trips away from my place. Our treks from one side to the other became a source of great

Above: Culvert construction across Rocky Creek
Below: Filling and compacting soil over the culvert

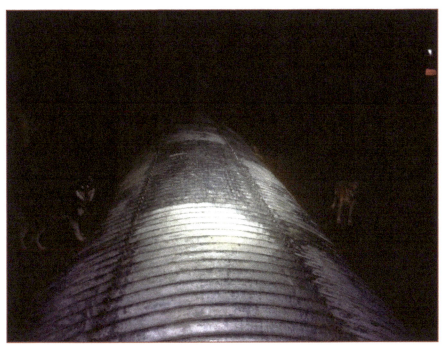

Above: A night walk across the construction area
Below: Concrete pouring

Above: Work continues
Below: Concrete forms removed

Above: Fill is almost complete
Below: Lots of fill to create a gentle enough slope for the road

entertainment. The dogs loved the routine, becoming familiar with the scents and signs along the way. And just as they were fully engaged in exploring the area we traversed, I loved this slowed down mode of travel. It was a time to reflect, to enjoy the gentle movement of the body, to breathe in deeply the moist forest air, to take in the subtle changes in the terrain and environment, to watch and listen for birds, to monitor culvert construction progress, and to see friends and neighbors.

I never knew who I might see along the way. It was, in almost all cases, a delight to get to say hello to people I otherwise would hardly ever see.

Those three months passed quickly. It was no surprise that something seemed to have been lost when construction was complete, and we were able to drive freely to and from our homes. Though a little inconvenient at times, most of us missed that slowed down commute.

15

National Resources Conservation Service Creates a Forest Management Grant

The days after the fire quickly turned into weeks and months. For most canyon residents, recovering from the trauma of the fire was not swift. Conversations with neighbors almost always moved to talking about the fire, even months afterwards. Everyone who lived in the Palo Colorado community had their own story to tell, with reflections upon the myriad ways things could have gone differently—for better or for worse.

In one of those conversations, someone mentioned that funding was available for managing forests burned by the fire.

I looked into this and soon found myself in a conversation with John Warner of National Resources Conservation Service (NRCS}. He told me of available funds for forest management projects and that my property just might be eligible.

We set up a meeting at their office in Salinas. I registered my property with United States Department of Agriculture (USDA) and then with NRCS and filled out an application for help with the burned forest. John came out to the property and took pictures. A few months later a state forester came for a site visit to create a forest management plan.

John told me to do nothing to the forest during the time the application was being processed, though I was tempted to pull up the blanket of young ceanothus trees coming up everywhere the fire had burned. After the winter and spring rains, the ashen grey earth on the northern slopes of the mountains gave way to a wild green bloom of plants, but predominantly ceanothus, their seeds having lain dormant in the earth for who knows how long.

The grant took a long time to secure, but when it finally came through in 2018, $57,000 had been allocated to manage 23 acres of burned forest.

Before any funds were sent, the work needed to be done, and the acreage where the work had been completed needed to be inspected. The allocation of funds was broken down into four payments, for four different areas on my property—one of eight acres and three of five acres each.

The designated tasks included removing, or clearing the brush through burning, mulching, or chipping, and then to remove 80% of dead standing trees. Five years later, the funding for the final five acres was being processed (August 24, 2023).

It was difficult, strenuous work. The terrain was steep, and the ceanothus had grown quite tall and impenetrably thick. Most of the ceanothus we burned during the winter 'burn season,' from the end of November to the end of April. Burning the brush had its challenges. Conditions needed to be suitable to work with fire. I preferred burning soon after a rain, when the ground and vegetation had a high moisture content. There were times I burned brush when it was dry, and was challenged more than once to keep the fire contained. The winter of 2021-22, I did no burning at all because conditions throughout that winter were simply too dry, and in my estimation, too risky.

That winter, on January 21, 2021, a canyon neighbor had a three-day old burn pile turn into a significant wildfire when high winds blew an ember into dry brush in close proximity. By the time the fire had been extinguished, 687 acres had burned. A little fire can become a big fire in no time at all.

It was good to have help when burning. It made the work more enjoyable and also much more was accomplished. Best was to gather a crew of six or eight people. On those days a significant area could be cleared and burned, while a sense of comradery enhanced the joy of the day.

Once a fire was started it was important to manage it carefully, while the bed of red-hot embers accumulated to a large enough body to sustain the fire. Once the ember bed was big enough, anything could be thrown onto the fire and it would burn—even green, freshly cut vegetation.

The most important thing was getting the fire out at the end of the day. Usually, we could rake all burnable material either away from or into the ember bed and leave it with a circle of bare soil around it, confident the fire had no fuel to begin to burn into the landscape. However, sometimes the wind would be blowing and to be safe, we covered the ember bed with a layer of earth.

Often, when we came back the next day, the same ember bed could be used to begin burning anew.

On a few different occasions, an ember lodged up in a cavity or nook of a dead tree and started a fire up in the tree. If the wind was blowing the fire would not easily go out but would slowly continue its burn. Sometimes these fires were 10-15 feet off the ground, necessitating cutting the dead tree down in order to douse the fire.

The most spectacular off the ground burn was in a hollow oak log. The tree had broken about 12 feet off the ground and the hollow trunk of the tree had fallen horizontally onto a steep hillside. The lower end of the hollow trunk was held up by the remainder of the still erect trunk, thus about 12 feet off the ground. An ember had lodged at one end of the hollow trunk and started the inside of the hollow trunk on fire. There was a slight breeze that night and air began to make its way through the hollow part of the trunk where the fire was blazing.

Someone from across the canyon had seen the fire in the night and called me. It was after 11pm. I immediately grabbed a chain saw and shovel and drove down to where the log was on fire. The fire sounded like a blow torch as the funneled air fueled the burn with plenty of oxygen. It looked like a huge wooden blow torch. I tried shoveling earth up into the trunk and onto the fire, but the fire would not go out. I had to cut the still standing part of the trunk down to get the rest of the tree fully onto the ground, and then spent the next hour fully extinguishing the smoldering part of the trunk.

Fortunately, most of the burn operations went smoothly, but working with fire in a wild landscape was unpredictable, making the endeavor naturally exhilarating. Every time when starting a fire, I felt a primal surge of attention and presence. Fire is a powerful force of nature and working in close proximity to it was energizing. I often found myself wishing that people separated from nature and the power of natural forces and phenomenon could experience what we did when doing the work in the forest—it created such a deep connection to the earth.

Once the brush was clear, bringing down the dead trees was easier. Trees would fall in the clear rather than into brush. There were, however, lots of dead trees. Steep hillsides made for slow, careful tree removal.

Ethan Atkins helping with firewood production

Steep hillsides to work on. Theo, Ethan Atkins, Brett Pallestrini, Asher Rose

Steep terrain to work the land

Firewood loaded and ready for delivery

Brush clearing crew – many hands make for light and enjoyable work

Wood splitting

Firewood production

Brush burning and firewood processing on a cold day are nice together

An incredible example of a spiral madrone. This one is worth saving.

Spiral madrone log makes it up to the main house.

The log that almost became firewood, becomes
a decorative art and functional installation.

It was dangerous work, especially when the top branches of trees were entangled. On more than one occasion, intending to bring one tree down, one further up the hill would come down with it because the branches were slightly entangled. After episodes with the unintended tree falling close to where I stood, I became ever more vigilant in scrutinizing every tree we brought down before initiating a cut.

Also, at the beginning of each day, I found it a good habit, to take a little time to come into a quiet place in the mind, in order to be fully present with the task at hand. It is never good to be in a hurry, and best to put attention on harmonizing one's efforts with the natural world. That always seems to bring about the best results. I often reminded myself and whoever was working with me that we were growing a forest, not trying to make money. The firewood business and grant were simply nice side benefits.

It was best not to damage the new trees growing from the old root balls, but was often difficult not to when bringing down the dead trees. Some of the new sprouts had diameters of over four inches and represented years of new growth. Through direct experience I learned ways of increasing the probability of dropping a tree where it would do the least amount of damage or fall completely in the clear. This could be done through the orientation of the initial pie cut, and then, how the cut from the other side was done to meet the pie cut. Depending on how that was done it was possible for trees to twist as they fell by cutting one side of the tree last. Because this section was still intact until the last, the tree would fall in this direction—pulled by the uncut grain of the tree.

Once a tree was on the ground it was cut into sections and then transported in one way or another to the nearest flat area for processing into firewood. Sometimes we rolled and threw the logs down the hills, and other times we used a truck or tractor with ropes, straps and chains to pull the wood to a workable area. That portion of the grant work, the removal of the dead trees, became a profitable firewood business, selling over 100 cords of firewood each year, from 2018-2023.

Some of the madrone trunks and branches were beautiful and unique. Madrone trees have a way of growing in spiral patterns or in unique curving arcs. Some of the trees were magnificent expressions of this and were so spectacular it did not seem right to cut them into firewood. There were also places along the trunk where two smaller trunks emerged, and once cut and turned upside down, made beautiful stools. The woodgrain of those parts of the trees also tended to be so intertwined that they could not be split into firewood. I collected these unique sections of the trees in hopes that at some point I would have time to process them into pieces people might want for landscaping or wood art, but currently most remain piled.

One log I almost cut into firewood became a beautiful artistic feature of my house. It was in a pile of wood we had rolled down the hill. As I started my first cut to turn it into rounds for firewood, I noticed it would be perfect for an entryway I was creating in my house. I immediately stopped the cut, turned it on its side and did a straight cut down the center of the log. The two finished halves were installed in the entryway for a beautiful effect.

Some other logs with extremely straight grain were milled by a friend into boards for a variety of uses. Though madrone wood has a reputation for twisting and warping once it is milled into boards, if you can find parts of the trees with straight grain, the wood is beautiful to use with its varied shades of brown.

I, and all the people who helped with the work in the forest, have grown to appreciate the beautiful expressions of natural art in the varied forms expressed by the trees. Every day we are in the forest we marvel at the wonders on display in the realms of nature.

Bleeding Heart – non native species in the landscaping around the house

Douglas Iris – native species to Long Ridge

Theo, open to the future

16

Living into the Present

Five years have passed since the first areas designated by the NRCS grant were cleared of brush and dead trees. Those areas have regrown into a diverse wildlife habitat, with a variety of species being expressed, instead of predominantly impenetrable ceanothus. I think the work has been a benefit to the biodiversity of the landscape, creating a mixed forest with wildlife corridors throughout. Wildlife populations in and around my property have not only rebounded but with many species, significantly increased compared to pre-Soberanes Fire conditions. In time, the effects, whether positive or negative, will become more apparent. Doing forest work like this is truly an experiment. One thing is certain—the areas we worked are more fire safe. If there is another fire it will be far less intense in the areas treated with brush clearance and dead tree removal.

The properties bordering mine look very different. On one border, nothing has been done. The ceanothus has grown into a dense thicket of trees 12-15 feet high with dead standing madrone intermingled within it. It is impenetrable unless you're willing to crawl on hands and knees and sometimes your belly. It is not pleasing to look at and it is a dangerous fire hazard. There is so much fuel. If there is another fire, this area will become an inferno generating intense heat and its own wind.

On the other side of my property a neighbor removed all shrubbery from the land and allowed grasses to grow. It is beautiful and very fire safe but is not so inviting to nesting birds or timid mammals.

After these many years on this plot of land in the Santa Lucia Mountains of Big Sur, I am left with more questions than answers. I wonder, what is our right relationship with the land? What are the best practices to enhance biodiversity and the health of the landscape? Is it better to do nothing at all? Are we meant to be living on the edge of or surrounded by wilderness? Can human presence and activity be

Theo's property – south and west facing hillside burned significantly

Two years later

The island of green in the midst of the burnt landscape

Two years later

beneficial to and enhance the environment supporting greater biodiversity and biomass?

My belief is that human presence across the landscapes of the world can be beneficial, but only if we honor our relationship to the rest of the life we share this planet with, and orient our daily lives and practices toward stewardship rather than dominance, conquest, and consumption of resources.

What does it take to be a good steward of the land and how can we orient ourselves toward earth-wise practices in the current human paradigm in which we live? How do we align our thoughts and actions with natural law, so that what we think and do benefits all of nature? These are the questions I live with daily.

The amount of destruction a raging fire can do to communities in a world of rapid climate change has been demonstrated in these recent years. The destruction of so many homes and buildings is a toxic burden to the environment, as well as a waste of precious resources and money. It is clear to me we are not living in accord with nature, or natural law.

In the aftermath of a major fire, it makes little sense to rebuild homes in the same places, and using the same construction techniques. From my experience there are ways to make your property, home, and buildings fire safe, and to enhance and enrich the environment. I think it is the responsible duty of everyone wishing to live in fire prone areas to do so.

One of the most interesting recent revelations in building is an invincibility effect that seems to be bestowed upon homes that are built according to Maharishi Vastu. Homes built in this way are oriented to the sun, the equator, and the earth's north-south axis. Rooms have their ideal placement in the house. There are precise dimensions for the design of rooms and the house as a whole. Slope of land is taken into consideration as well as proximity and orientation to bodies of water. It is ancient wisdom of natural law put into practice when a home or building is constructed that is for work, or human habitation. These buildings are built in harmony with natural law, and it seems natural disasters spare these homes. In more than one instance, fires have gone around houses built according to Maharishi Vastu principles, leaving them miraculously unscathed—what is now being called, The Maharishi Vastu invincibility effect.

When I built my home, I integrated into its design and construction as many vastu principles as I knew, and the Soberanes Fire did seem to sweep right around all my structures.

Our accepted way of living is rendering the planet ever less habitable and, all the while, most people do nothing to change their habits or lifestyles to minimize or mitigate the destructive results of their choices.

Simultaneously, we have put in place around the globe, educational curriculums with an emphasis on subjects and experiences that do nothing to help people live in harmony with the laws of nature, support individual awakening, or preserve life on the planet.

It is clear to me that we are in this life experience together. I am not only speaking of my ridge top community, but the entire interconnected web of life. We are one human family and one symphony of life. It is in the strength of our shared vision and energy that the most beautiful world can be realized, as well as the most fulfilling lives lived.

In the final analysis, it is this sense of unity that could and hopefully will help create the shifts necessary to alter the destructive path our lifestyle choices are creating on this beautiful planet. Have we not seen enough destruction wrought by our greed and senseless strivings? Would we not be so much better off acknowledging our interdependence and doing everything we are capable of to strengthen our collaborative efforts?

I can only imagine the amazingly beautiful world we could create if we put our shared attention on stewarding ourselves and this lively planet earth.

We are the only living creature on the planet that can respond to the call to stewardship. We are the ones who can gain a sensitivity to the needs of the world around us and to do what is necessary to enhance life, rather than diminish and take from it. We are the ones who can understand the subtle laws of nature and align our efforts with those laws to establish a true heaven on earth.

Separating ourselves off in human-made environments may seem safer than immersing oneself in the ever-changing landscape, but I can assure you that it is in the ever-changing landscape that the universe's secrets are revealed, the best food is found, the coolest things are seen, heard, and smelled, and life can be lived to the fullest.

Nature is my favorite teacher. When we open ourselves wholly and fearlessly, we can experience an enlightening guidance, one that is beyond our normal thinking and even beyond a sense of self. Direct experience becomes presence. Presence allows us to attain the open-hearted sensitivity in which we can feel our connection to everything around us, creating the opportunity to experience the fullness of life, and the true understanding of our interconnectedness. Presence allows us to know ourselves and to understand what we are being asked to do next, not only for our individual best interests, but for the best interests of everything else as well.

What better medicine is there than to get outside and follow nature's lead?

Azu and Osiris, faithful companions along the way

17

House and Land Performance

When I first bought the property in the Santa Lucia mountains, I spent months becoming sensitive to the land that comprised it. The intention of that expenditure of time was to align my activities on the land in order to nurture and enhance the life here through sensitive and gentle Earth-wise practices. For the ten years preceding the Soberanes Fire, I prepared for what had always seemed inevitable—fire. I think that preparation not only made the landscape healthier, but also made all the difference in making it through the fire with structures and landscaping intact.

The way the buildings were built, and the way the land was tended, created a situation that worked to keep the fire from burning the structures, and the vegetation close to the structures. The area around the structures provided a safe zone which CalFire used as a base of operations while protecting the homes on the upper part of Long Ridge. The island of green, unburned landscape became a refuge for an abundance of local wildlife.

The Rain Catch System

- The water system performed flawlessly, until the water timers were burned during the backfire operations, allowing the water left in two tanks to drain completely. Until then, the system provided enough water for firefighters to fight the fire on Long Ridge. CalFire used the system to refill their engines and to back burn around nine houses on the ridge. CalFire also laid over 900 feet of hose from one of the hydrants in the water system in an attempt to save a nearby house.
- Having a 500-gallon water trailer with a pump and fire hoses was a good thing. It was useful in extinguishing flames and providing water under pressure, up to a distance of 75 feet.

- Buried pvc water lines kept them from burning.
- Multiple hydrants allowed fire fighters to easily hook up multiple hoses creating a widely defensible situation.
- A valve and outlet on the main road allowed fire engines to fill from storage tanks.
- Over 35,000 gallons of water storage at the start of the fire gave CalFire the needed resource to do the back burning around houses. The total water storage capacity is close to 60,000 gallons. All the water is collected during the rainy season, which every year has filled the tanks completely. As the dry months pass, the water is used mostly for irrigation, leaving between 10 and 20 thousand gallons once the rains begin again in the fall.

Buildings

The house and buildings did not burn. Building with fire resistant materials made the difference between losing structures and coming through with all of them intact. Even when the leaves in the gutter of the barn caught fire, the structure did not ignite. Those leaves burned for at least ten minutes before we arrived on scene, but the flames were contained in fire resistant materials. As a result, we were able to extinguish the flames in a few minutes.

Cold Room/Fire Shelter

The cold room worked more wonderfully than I could have imagined. The hours spent sleeping in that little room were as nice as any hotel room. The compressed air gently escaping from the SCUBA tank was cool and sweet and created a positive pressure, keeping the smoke-filled air outside from entering. As the air escaped from the opened tank, ice formed on the tank, cooling the room to a very pleasant temperature.

The cold room/fire shelter provided a safe, smoke-free place to stay on the land and manage the fire. I was quite certain, that even if the house and shop burned, I could survive in the cold room/fire shelter for many hours. Without the shelter, I would have been reluctant to stay.

Anyone living rurally in a fire prone landscape should put the energy into creating a fire shelter. Evacuation is not always an option in a raging wildfire and roads can become impassable, especially with one road for both ingress and egress.

Perhaps most importantly, a fire shelter provides the safety needed to support the decision to stay and defend structures, animals, and landscaping.

Brush Clearing, Fire Mimicry

Years were spent preparing for the Soberanes Fire. I bought the land knowing that fire was the biggest threat. Building construction was carried out with that in mind and time was spent every year making the space around buildings increasingly more fire safe.

The many years of brush clearing/fire mimicry stopped the fire's advance. When the fire began to burn across the land where the brush had been extensively cleared, it either stopped altogether or slowed to such an extent that it was manageable and easy to put out using the simplest of fire-fighting practices—shoveling dirt onto the fire and creating a fire break of bare mineral soil.

Being able to bring the fire to a standstill Saturday night and Sunday morning provided the necessary window of time for CalFire to arrive, whose efforts in defending Long Ridge saved people's homes.

Firefighting Training

Though becoming a firefighter is not a feasible option for many people, I cannot express strongly enough the value of the knowledge and experience gained during the six years I spent serving the Palo Colorado and Big Sur communities as a member of Mid-Coast Fire Brigade. Not only did those years forge a deeper connection with community, but also gave me the confidence, experience, and skills to stay and deal with the fire, rather than evacuate. If I had evacuated, there is little doubt in my mind that my neighbors and I would have lost everything.

- Get Firefighting Training
- Watch YouTube Videos on firefighting
- Read books about firefighting successes and failures
- Get proper firefighting apparel and tools

Build According to Maharishi Vastu

I think it is worth looking into the unusual and powerful benefits of building according to Maharishi Vastu. There are numerous accounts of fires and other natural disasters going around buildings built according to the natural laws employed in Maharishi Vastu. It is known as the Maharishi Vastu Effect or the Invincibility Effect. The Maharishi Vastu Effect seems to prevent negative influences damaging a home built in this way. www.maharishivastu.org

Considerations for Building and Living in Fire Prone areas

Below is a list of things to consider when building and/or living in an area subject to regular fires.

- Choose fire-proof building materials for all exterior surfaces—concrete board, tile, stucco, concrete, steel.
- Cover all soffits with a fireproof material.
- Consider installing storm shutters.
- Enclose the undersides of all buildings so a burning ember cannot blow beneath the building and ignite the structure.
- Build a water system with hydrants and fittings that enable fire services to connect.
- Bury water lines so they do not burn/melt in the fire.
- Keep flammable brush and debris away from water tanks.
- Place propane tanks on a concrete pad or equivalent, away from buildings, and clear brush in close proximity to the propane tanks.
- Have a way for your water system to work for you during an emergency—a system that will not burn or be affected by fire—at least one metal water tank and buried waterlines.
- Become a member of your local Fire Brigade.
- Get firefighting training.
- Have your own firefighting apparatus—hoses, nozzles, water tender/trailer, pump to boost pressure.

- Have firefighting gear—fire-rated jacket, pants, helmet, shroud, gloves, boots (cotton will not burn, though most synthetic fabrics do). Heavy duty, 100% cotton denim is a good choice if fire rated apparel is unavailable. Having a respirator and goggles is also a must.
- Have fire-fighting tools—shovels, pick axes, Pulaski axe, McLeod
- Learn how to use a shovel to fight fires effectively—the shovel is perhaps your most important tool.
- Create significant brush clearance around all structures—a minimum of 100 feet.
- Have a perimeter of bare mineral soil or the equivalent around your home and buildings.
- Have a place of refuge—a cold room or bunker, something totally fire-proof and preferably underground. Inside your shelter have food and water, a place to rest and preferably lay down, a clean air source with an air filter, or a SCUBA tank with air, and extra clothes.

Build Community

Having good relationships with neighbors and community members can be of great importance during an emergency. When a catastrophe occurs, having others to stand with you, and together handle the situation is generally far more effective than handling it by your self.

Have a Plan

Of great importance is knowing what to do when there is a fire, and to have defined a strategy to deal with it—long before the fire event happens.

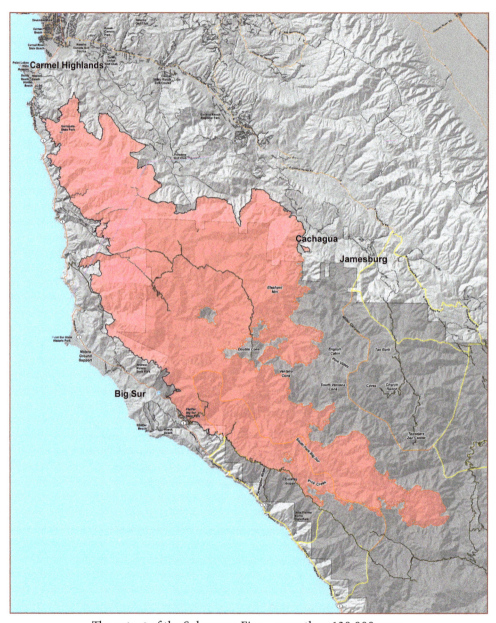

The extent of the Soberanes Fire – more than 130,000 acres

18

Making Your Land Fire-Safe

What does it take to live safely in a fire prone landscape? It takes conscious and focused effort, getting to know the landscape, and ideally, training in fire safety, firefighting, and fire suppression.
Fire needs three things to begin: an ignition source, oxygen, and fuel (a combustible material).

Fire is defined as chemical reaction: the rapid oxidation of a material which releases heat, light, and various other reaction products depending on the material combusting.

Fire is a fascinating, dynamic element. It has a quality making it seem alive. Used with care it can be an incredible tool, but when mishandled or not intelligently managed, fire can be life threatening and dangerous.

As a child, growing up on a farm in the Ohio countryside, it was my job and that of my two older brothers to burn the trash in a 55-gallon metal drum. We burned everything from paper to plastic. It was interesting to see the different colors of the flame, and to smell the different smells, depending on what was burning. I found fire to be a captivating force of nature.

Every structure fire I have ever been close to emits the most wretched smells as so many human-made products burn. When dealing with structure fires as a firefighter it was essential to wear a SCBA (self-contained breathing apparatus) to stay safe. Even with that precaution, it would take a few days to recover from inhaling smoke and fumes. Structures also concentrate fuels, and typically house substances are extremely combustible. Most structures once on fire are difficult to stop burning because of this. The fire goes out when the fuel is exhausted, or the oxidation process extinguished.

Wildfire depends on fuel to burn. The greater amount of fuel, the bigger and hotter the fire will be. The bigger and hotter the fire, the more easily it can maintain itself through the landscape.

Brush clearing is an attempt to mimic fire behavior and create a landscape where fire will not be able to get too hot or too big to be unmanageable.

There is great debate over how to manage our landscapes, and where it is beneficial to use fire to do so.

It is clear when we let fuel loads build up too much, it creates dangerous and destructive fires. Drought conditions only make the potential for a catastrophic fire even more likely. Whether created naturally or by anthropogenic influence, it seems weather patterns are changing and less predictable. The earth seems more hot, more dry, and less hospitable.

One thing is certain - where there is no fuel, there is no fire.

In December 2014, I responded to the Pfeiffer Fire. Arriving on scene a little after midnight, we were assigned to provide protection for a dwelling along Pfeiffer Ridge Road. Having done everything possible to stop the fire's advance across the hillside below the house, the instruction was to gather hoses and carry them up the steep hill and back to the engine parked in the driveway. Just as the darkness of the night was being dispelled by morning light, there was a wind shift. Somewhere below us the fire intensified, gathering a frightful momentum as the wind drove the fire uphill through dense patches of dry redwood duff and dead fallen branches.

There were three of us bringing up the gear, and I was the lowest. Suddenly, our chief instructed us to drop the gear and run to safety—run!! I had never been given those instructions before.

It was still a long way from where the land had been leveled out to build the house, and it was clear I could not outrun the fire. The two firefighters above me had no trouble reaching that safe level ground.

Having surveyed the landscape while climbing the hill, to the right, about 60 feet away was an area that had already burned. Areas like that are called 'black' because they are predominantly made of the charred remains from the fire.

After dropping the hoses I was lugging up the hill, I turned and ran toward the 'black.' The first few strides, I wondered whether I would make it or be overtaken by the fire. I had looked back for a moment when we were given that initial instruction to run, and was amazed at how quickly the intense flames were moving. What was more intimidating was the noise. The fire sounded like something between a jet engine and a train, and was becoming louder as it came closer. As it advanced up the hill, it was also creating its own wind, pulling the cooler oxygen rich air toward the rapidly combusting fuel.

I ran, feeling the heat from the fire through my wildland gear, and especially through the fire shroud that covered my neck. Making it into the area that had already burned, I dropped to the ground and looked back. The fire was intensely burning the area where I had dropped the fire hoses. Those first few steps had kept me from burning up with those hoses.

By the time I was in the 'black,' the fire had burned to the vegetation and accumulated dead material directly below the driveway where the engine was parked. Already, our fire chief had a fire hose deployed, directing the stream of water at the flames. I drew in fresh air with my face close to the ground, and as soon as possible, made it up to join the rest of the crew.

Later, returning to the fire station after 52 hours deployed, I inspected my gear and found that the back of my headlamp had partly melted from the heat of that close encounter with the fire.

I was thankful for that patch of already burned landscape.

In fire mimicry, we want to create a fire safe zone just like that, and reduce, or better yet eliminate fuel—combustible material.

One of the best tools to manage and stop a fire is a shovel. If a ground fire comes to a place where there is nothing but bare mineral soil, it cannot continue. Creating a line of bare mineral soil is a common technique used when firefighting—done on a small scale alone, with a hand crew and hand tools, or on a larger scale by earth movers like a bulldozer.

Bare mineral soil devoid of organic matter will not burn in a wild environment. In and around the home, concrete or stone hardscaping provides the same effect.

Most of the buildings we create we do so with the hope they will last through time. Many would say it is folly to build homes in wild places, but if built intelligently, it is a beautiful opportunity to harmonize our life and efforts with the land and the life upon it.

Things in the environment are always changing and as they do we need to diligently maintain the safe zone around our created environment. This usually means spending time each year maintaining the fire safe zone, and as much as possible, mimicking fire.

Some think it best to leave nature alone, but I believe we can adopt the role of stewards, and in the process enhance the land's productivity and biodiversity. Though it is not possible to go back in time to see, I think indigenous cultures did just that for centuries if not millenniums. Familiarity with the landscape is a must, which means getting outside and into the natural world. Knowing plant and animal species is of critical importance, to at least begin to make informed decisions about what to nurture and what to eliminate.

On my property, I began clearing brush on the south facing slopes—primarily chamise and dead manzanita branches. Each of these species have sizeable woody root balls, enabling them to remain green and growing through the dry months. The root balls also become large enough to survive through fire. Though the top and exposed parts of the plant will burn away, within a short time new shoots appear, brilliantly green against the blackened root balls.

Because of their ability to survive through fire episodes, I have wondered how old the individual plants actually are, especially the manzanita root balls which can be 3-4 feet in diameter. Certainly, some of them are hundreds of years old, but could they be thousands of years old? These ancient beings and their adaptive life cycle have earned my respect. As a result, it's best to minimally alter the landscape to create the desired effect.

In Big Sur, burn season, as we call it, begins when it is safe to burn brush in the fall, and continues usually through April 30th. Most years I take advantage of this opportunity, clearing and burning brush to maintain and sometimes extend the safe zone around buildings and landscaping.

Experience has taught that it is best to begin by starting a fire with a small amount of fine material and then add larger material little by little, enhancing the fire until it can burn anything thrown into it. I have found it dangerous to have a large pile of combustible material and to start the whole thing ablaze, as it can turn into an unexpectedly large fire. This can lead to an out-of-control situation—a highly unfavorable and potentially expensive outcome.

When burning brush, it is good to have a clear area to burn in, making sure the fire is not too close to other standing, uncut brush or trees. Use a shovel to create a perimeter of bare soil around the burn zone. It is important to have tools close by to work the fire (shovels, pick-axes, McLeod, Pulaski), a water source handy, and the time to stay with the fire until it is out.

On January 21, 2022, the Colorado Fire in Big Sur was ignited by a 3-day old burn pile. High winds blew hot buried embers into surrounding brush resulting in almost 700 acres burned. The moral of this story, if conditions are right, fire can start in California even in the winter months. Make certain your fire is completely out! That fall and winter were very dry. As a result, I did not do any brush burning that year—it just felt dangerously dry. Wisely choosing when and when not to burn is essential. Making the right choice begins with common sense, and with experience evolves into a well-informed decision.

It is amazing that once there is a good blaze going, freshly cut green trees and brush burn readily. Not only does it make the job of fire mimicry on the landscape easy, but makes it clear that fire is a powerful natural force to be treated and used with the utmost respect.

There are alternatives to burning brush, including mulching, chipping, or burying vegetative material. The advantage to any of these other practices is that they can be done any time of the year.

My property straddles a ridge. Down the south side of the ridge is Turner Creek and down the north side is Rocky Creek. In California, steep north slopes tend to be shady, supporting tree growth and forests. Steep south facing slopes are typically made up of drought tolerant grasses, shrubs and stunted trees—also known as chapparal.

In the parts of the property that are oriented toward the south, receiving high sun exposure, there is a lot of manzanita and chamise. manzanita is a low growing shrub. The woody part of manzanita is very dense, and though it will burn when it is alive, it does not burn easily. The bark is a dark brownish red, and it has small green leaves present year-round. chamise is a brushy shrub. It does not have leaves but instead bristly branches and stalks. It is dry, grows densely over the landscape, and is difficult to get through. It readily burns.

On those south facing slopes, I remove dead manzanita branches and as much chamise as possible. Most often, I burn as I clear. It has never worked for me to make piles with the hope of coming back later to burn. I sometimes would not make it back to do that and even after only a couple days, brush can dry out to the point where it burns hotter and faster.

The end result of doing fire mimicry work is a much more open landscape with islands of manzanita. Having islands of vegetation make it much more difficult for a fire to make progress across the landscape—the vegetation is not continuous, nor contiguous. This also makes it easier for wildlife to move in and around the area.

When the Soberanes Fire reached the south facing areas where the chamise had been cleared, it was unable to proceed across bare soil to the next brushy area, or island of manzanita. As a result, in those areas, the fire stopped burning as it moved up hill toward the gardens, orchards, and buildings.

On the north slopes, within a beautiful, 50-year-old forest, there was a variety of mature trees, made up mostly of madrone, but also bay laurel, black oak, and tan bark oak. In the years before the Soberanes Fire, time had been spent going through many acres of this forest, cutting down dead trees and branches. Some of this wood was cut and split and used to heat the house. Most of the wood was for firewood sales—usually 10-12 cords each year.

When the Soberanes Fire came through the forest, it burned the many inches of leaves and organic matter that had collected on the ground through the years. Any dead branches in the path of the fire increased the fire's intensity. By the time the fire had swept through the organic matter on top of and in the soil, the level of the ground had dropped by 4-7 inches, and what was left had very little or no organic matter in it.

Madrone trees are the most common species of tree in the forest which makes up the majority of my property. I find them the most beautiful and whimsical of trees. They twist and spiral and grow into each other. They grow into the most amazing shapes. They generally have very thin bark, which in most places is more like a smooth skin than a bark. Beneath this smooth skin is the cambium layer—the part within the trunk and branches of the tree that is alive. The thin bark makes the cambium layer of madrone trees very susceptible to fire. Though the entire tree does not easily burn, once the cambium is burned, everything above it dies. This was true of 80% of the madrone trees in the parts of the forest that burned.

That next spring, after the fire, it was amazing to see the top parts of partially burned trees still alive. Those trees stayed alive long enough to put out the greatest outpouring of blossoms any of us in the area had ever seen. What was left of the madrone canopy turned into the most impressive display of white, due to the density of the blossoms. Those blossoms then turned into seed carrying berries that littered the ground in the fall, seeding the landscape for a new generation of madrone trees. The intelligence exhibited by these trees in dealing with the trauma of fire and doing everything possible to perpetuate themselves is impressive.

Once the top part of the madrones died, the roots/base of the trees sent out new green shoots, starting the growth process toward trees once again.

It turns out the old dead trunks hold an amazing amount of moisture, providing water to those new sprouts during the drier months. When these dead trunks are cut down, even 6 years later, the bottom 10+ feet is often still very wet—yet another example of intelligent natural design at work. How many times have these trees been burned by fire and come back?

This same quality of regrowth from the roots holds true for many of the species of shrub and tree in the Big Sur coastal landscape. Redwoods, black oaks, bay laurel, manzanita, tan bark oak, and coast live oak, all grow back from their roots. How old are they?

What would fire mimicry on a large scale look like?

It seems the indigenous Californians used fire to keep the landscape vibrant and open. Regular burns across the landscape kept those areas free of low growing plants, making the landscape more passable for both animals and people. Large mammals thrived, and hunting was easier. It is challenging to mimic their techniques in the world we live in today because of our 'permanent' building practices. They did burns on a large scale and much more regularly. As a result of the regularity of the burns, the fuel loads never built up to the amounts we often end up with today. Excessive fuel loads, once a fire ignites, result in the mega-fires we more commonly experience these days.

Apparently, the indigenous people in our area used fire to clear the landscape annually and sometimes even bi-annually. Purportedly, the regular burns kept the forests open and clear, kept brushy fuels from getting too big, allowed many species that needed fire to flourish, and overall made for a vibrant, healthy ecosystem. Regular burns keep fuel loads at a minimum, eliminating the threat of huge disastrous fires that alter the landscape and wildlife populations for years.

What would it take to replicate landscapes like the ones maintained by indigenous cultures?

Individual landowners would need to maintain their own properties with fire mimicry practices, while governmental agencies would need to commit funds and resources toward fire safe land management practices on publicly held lands. There is however, great debate about the most effective land management techniques, as well as whether human manipulation of the environment can have any beneficial effects at all.

It would be interesting to compare the cost of dealing with a raging wildfire and the destruction it causes, to the expense of maintaining fire safe landscapes.

Dealing with raging out of control wildfires is costly, and the aftermath is as well. The total cost for resources called upon in an attempt to manage a fire is akin to the cost of war: air support, engines of all sizes and configurations, water tenders, bulldozers, excavators, ambulances, and hundreds or thousands of fire fighters. Entire temporary villages are put in place to house and take care of personnel. The sad truth is that every year, inevitably lives are lost. In the aftermath, the landscape is forever altered. Wildlife populations may take years to rebound. Property owners who might have lost everything must decide how to proceed. Sometimes entire communities are devastated, as was the case in Paradise, California in 2018. The cost to the environment, which we don't measure, is grand and far reaching. While human-made materials burn, the atmosphere becomes the recipient of toxins which later rain down upon the landscape, often thousands of miles away. To rebuild or reconstruct means resources once in place and in use, must be gathered and assembled again, placing an unnecessary burden on an already maxed out supply chain dependent on the limited resources of the planet.

The Soberanes Fire, which burned 130,000 acres, cost $230,000,000 in resources, and took 82 days to put out (clearly, the "We've got it handled" comment by the IC that first day was far from true). If you look to statistics for dollars spent each year to battle wildfires and the cost of the aftermath, in California alone, it is an excessively high amount—in 2018 the cost to combat California Wildfires was over 145 billion, according to University College London. Nationwide, that figure is certainly much higher.

How can we change this paradigm? That same amount of money used to establish fire mimicry practices on public lands and assist private landowners to establish those practices on their lands, might be money far better spent. This process would have multiple benefits—enhancing our life-giving planet, giving meaningful employment to millions, and educating and informing a new generation of land stewards.

To me, the most reasonable path into the future is to steward some landscapes using all the fire safe, land management techniques available, in an attempt to create more vibrant ecosystems, and enlightened human beings in the process. Where is the enlightened leadership to create and put into place the programs necessary?

If we do not choose to steward this beautiful planet, the alternative is to continue to look for solace in a materialistic focus that only brings about the continued destruction of the life-giving processes of this planet. We either sit idly by tending to things of little importance or take our place in concert with the whole, dropping the I and claiming the We.

The choice remains ours. Do we sit back and watch as negative influences continue to ravage our world, or do we take an active stance as stewards and join in the co-creation of a vibrant, safe, and life-filled world?

Theo in his natural environment

Acknowledgements

As I reflect on everything that has occurred to bring this book into print in this form, there are so many people to thank and acknowledge.

It may have started with Kim Bancroft, from both my high school and college days, and an author, who kept encouraging me to keep writing and put me in touch with Marthine Satris, Acquisitions Editor at Heyday Press. I sent Marthine a manuscript that included my story of the Soberanes Fire. She gave me the constructive feedback that I was too all over the place in my writing. I took that manuscript and divided it into four different books, this one being the first and I think the most important, given the out-of-control fire seasons we are experiencing in the western United States.

Then there's Jake and Cheryl Goetz and Brent Bispo, who oversaw my introduction to wildfires and my safe relationship to them during my time serving as a firefighter on Mid Coast Fire Brigade. Had it not been for Brent, Jake, and Cheryl's tutelage, I would probably not have a home to tell this story about.

My son, Chris Maehr, showed up to help when things were the most intense during the Soberanes Fire. He made all the difference for me being able to safely stay on my property and gave me the support I needed to keep enduring through feverish states and extreme conditions. His company and willingness to do whatever needed to be done I am forever grateful for.

My sweet dog Azu kept me company through the whole ordeal, and licked me back into consciousness when I feinted turning off the generator. She was continual comfort always letting me know I was not alone on the mountain.

Kierstyn Berlin shared her jubilant energy, provided a safe shelter away from the fire, and a place to recover once I left the mountain.

Maria Laura Ortega, my beloved while creating this book, helped me realize the importance of never letting anything stand in the way of unfolding your dharma. As the Bhagavad Gita so illustrates, it is most important to fulfill one's duties in accordance with one's nature. It is better to perform one's own duties imperfectly than to master the duties of another. By fulfilling the obligations one is born with, a person never comes to grief.

Ginna and David Gordon made this book a reality, taking my manuscript and pictures and sculpting them into the volume you have before you. Their support and encouragement kept me focused on every minute detail. Their ability to navigate through the publication process is, to me, miraculous.

Travis Trapkus never stopped encouraging me to stay on my property until everything was safe. He showed up often to check in on me to make sure I was OK, and came to help when it was time to get the horses to a safe place. It was also Travis and Brent Bispo who shared their construction know-how and helped me create the fire shelter, without which I would have been reluctant to stay and defend Long Ridge.

Patricia Bercovich, checked in with me daily, and kept friends and family appraised of my status with the fire. Patricia spent four years living with me on the property and contributing in ways for which I am forever grateful. She purchased the water trailer one especially dry winter and helped fund the building of the fire shelter/cold room. Her friendship and support have endured through many years, even up to the final read through edit of this book.

Zach Esbach, and Pat and Brett Palistrini kept a watchful eye on Long Ridge and my property during the times when I was not there, and also kept patrolling and managing fires on Long Ridge until things were safe.

The many Mid-Coast Fire Brigade and CalFire firefighters, who tirelessly engaged with the fire to protect the Palo Colorado community.

Todd Champagne created the birthday event that surrounded me with community and friends, helping me integrate all that had gone

About the Author

Theo Maehr believes that in living closely with the land we can unfold our full human potential and better understand our connections with everything and everyone. Having chosen to find his wealth in the abundance nature provides, he has lived in the wilds of Big Sur for over 20 years.

In addition to being an author, Theo is an educator, lyre maker, house builder, fine wood-worker, diver, nature enthusiast, and steward of the land.

His prior endeavors include Waldorf teaching, home-school oversight, professional storytelling, outdoor education, agriculture, construction, raft guiding, and lots of time wandering around in the wilds.

Theo currently lives in the Santa Lucia mountains of Big Sur, California, in the sturdy, off the grid home he kept from burning in the Soberanes fire. He holds a Bachelor of Science in Geologic and Environmental Science with an emphasis in Land Resources, and a Master of Arts in Education with an emphasis in Curriculum Design and Teacher Training, both from Stanford University. He is available as a consultant, to share his wisdom of land stewardship and fire mitigation.

Visit Theo's website: www.theearthsteward.com

Printed in the USA
CPSIA information can be obtained
at www.ICGtesting.com
LVHW072041050824
787263LV00014B/257